IN VIVO

The Cultural Mediations of Biomedical Science

PHILLIP THURTLE and ROBERT MITCHELL, Series Editors

IN VIVO: THE CULTURAL MEDIATIONS OF BIOMEDICAL SCIENCE
is dedicated to the interdisciplinary study of the medical and life sciences,
with a focus on the scientific and cultural practices used to process data,
model knowledge, and communicate about biomedical science. Through
historical, artistic, media, social, and literary analysis, books in the series
seek to understand and explain the key conceptual issues that animate and
inform biomedical developments.

The Transparent Body: A Cultural Analysis of Medical Imaging
by José Van Dijck

Generating Bodies and Gendered Selves:
The Rhetoric of Reproduction in Early Modern England
by Eve Keller

The Emergence of Genetic Rationality: Space, Time, and Information in American
Biological Science, 1870–1920
by Phillip Thurtle

Bits of Life: Feminist Studies of Media, Biocultures, and Technoscience
edited by Anneke Smelik and Nina Lykke

Life as Surplus: Biotechnology and Capitalism in the Neoliberal Era
by Melinda Cooper

HIV Interventions: Biomedicine and the Traffic between Information and Flesh
by Marsha Rosengarten

Bioart and the Vitality of Media
by Robert Mitchell

Affect and Artificial Intelligence
by Elizabeth A. Wilson

Darwin's Pharmacy: Sex, Plants, and the Evolution of the Noösphere
by Richard M. Doyle

The Clinic and Elsewhere: Addiction, Adolescents, and the Afterlife of Therapy
by Todd Meyers

The Pulse of Modernism: Physiological Aesthetics in Fin-de-Siècle Europe
by Robert Michael Brain

Tracing Autism: Uncertainty, Ambiguity, and the Affective Labor of Neuroscience
by Des Fitzgerald

TRACING AUTISM

Uncertainty, Ambiguity, and the Affective Labor of Neuroscience

Des Fitzgerald

UNIVERSITY OF
WASHINGTON PRESS
Seattle and London

Sponsored in part by Duke University's Center for Interdisciplinary Studies in Science and Cultural Theory

UNIVERSITY OF WASHINGTON PRESS
www.washington.edu/uwpress

LIBRARY OF CONGRESS CATALOGING-IN-PUBLICATION DATA
Names: Fitzgerald, Des, author
Title: Tracing autism : uncertainty, ambiguity, and the affective labor of neuroscience / Des Fitzgerald.
Description: Seattle : University of Washington Press, [2017] | Series: In vivo : the cultural mediations of biomedical science | Includes bibliographical references and index.
Identifiers: LCCN 2016049325| ISBN 9780295741901 (hardcover : alk. paper) | ISBN 9780295741918 (pbk. : alk. paper)
Subjects: | MESH: Autistic Disorder—diagnosis | Neurosciences
Classification: LCC RC553.A88 | NLM WS 350.8.P4 | DDC 616.85/882—dc23
LC record available at https://lccn.loc.gov/2016049325

For Anne Fitzgerald and Tom Fitzgerald

It is possible that the very productive critical habits embodied in what Paul Ricoeur memorably called "the hermeneutics of suspicion"—widespread critical habits indeed, perhaps by now nearly synonymous with criticism itself—may have had an unintentionally stultifying side-effect: they may have made it less rather than more possible to unpack the local, contingent relations between any piece of knowledge and its narrative/epistemological entailments for the seeker, knower, or teller.

—EVE KOSOFSKY SEDGWICK, "Paranoid Reading and Reparative Reading, or, You're So Paranoid, You Probably Think This Essay Is about You," in *Touching Feeling: Affect, Pedagogy, Performativity*

Contents

Acknowledgments

ONE OF THE MANY DRAWBACKS OF DELAYING THE PRODUCTION OF A manuscript for several years is that you build up a weight of debt and obligation that far, far exceeds what a typical acknowledgments section can bear. What follows is probably more a record of forgetfulness than it is of thanks. Oh well. First, I want to acknowledge everyone who consented to be interviewed for this book; that the book is partly an attempt to think seriously with and about intellectual generosity has much to do with the patience and care of those anonymous interviewees.

The writing of this book took place over two main periods: it began around 2010–11 at the BIOS Centre, then part of the Department of Sociology at the London School of Economics. I gratefully acknowledge financial support from that department. At BIOS the gradual unfolding of this manuscript would not have happened without the incredibly generous and careful support of Nikolas Rose; indeed, the manuscript would not exist in any vaguely readable or interesting form without his interventions. The extent to which I gesture at a twenty-first-century sociology, and how that discipline might relate to the biosciences, is largely derivative of things I have learned—and continue to learn—from Nikolas. During this period and beyond, Ilina Singh has been one of the most nuanced people I know thinking across the lines of the social, the psychological, and the ethical, and I have been very lucky to think and coauthor with her.

I am especially grateful for the guidance, friendship, and intellectual support of, also at BIOS, Joelle Abi-Rached, Megan Clinch, Victoria Dyas, Caitlin Cockerton, Caitlin Connors, Sabrina Fernandez, Angela Filipe, Susanna Finlay, Giovanni Frazzetto, Carrie Friese, Cathy Herbrand, John MacArtney, Maurizio Meloni, Sara Tocchetti, Scott Vrecko, Ayo Wahlberg, and especially Amy Hinterberger, who provided vital feedback on different parts of this manuscript. No doubt, there are many others who have been brushed from my memory by the onset of early middle age. In the sociology department at the London School of Economics, I learned a great deal from being around the writing of Malcolm James, Juljan Krause, Naaz Rashid, Jill Timms (and many others) as well as everyone at NYLON (a PhD writing group working between New York and London), particularly Adam Kaasa.

The second period of writing was during 2015 at the School of Social Sciences, Cardiff University, where I completed this manuscript under the immense generosity of an early career lectureship. I am especially grateful to have found myself among a brilliant collection of thinkers on science, medicine, and related topics. I have been especially glad of interactions with Michael Arribas-Ayllon, Paul Atkinson, Beck Dimond, David Frayne, Carina Girvan, Tom Hall, William Housley, Adam Hedgecoe, Kath Job, Joanna Latimer, Jamie Lewis, Sara MacBride-Stewart, Kate Moles, Hannah O'Mahoney, Robin Smith, Tom Slater, Steven Stanley, Gareth Thomas, Martin Weinel, Emilie Whitaker, and everyone at the research group on Medicine, Society, and Culture (MESC) as well as those within the research group on Knowledge, Expertise, and Society (KES), who commented extensively on the final chapter. On this last score I am especially grateful to Harry Collins, Rob Evans, and Sara Delamont.

In between these two periods of writing, I moved about as a researcher on various projects that involved thinking with and through the neurosciences. Although I didn't directly work on this book during these years, the people I met, spoke to, and worked with remain intensely present in the manuscript. I learned a lot about neuroscience and interaction during a nine-month stay at the Interacting Minds Centre, Aarhus University. This book owes a huge amount to the generous fellowship of Andreas Roepstorff, Uffe Juul Jensen, Joshua Skewes, and Svenja Matusall. I also learned a lot about collaboration while working as a postdoc (supported by the Economic and Social Research Council) in what is now the Department of Global Health and Social Medicine at King's College London. In addition to many of the BIOS personnel listed earlier, I salute Paula Bello, Silvia Camporesi, Giorgio Di Gessa, Federica Lucivero (all the Italians basically), Christine Aicardi, Orkideh Behrouzan, Carlo Caduff, Valerie D'Astous, Guntars Ermansons, Hannah Kienzler, Sam Maclean, Tara Mahfoud, Francisco Ortega, Barbara Prainsack, Michael Reinsborough, Sebastian Rojas, everyone at the Culture, Medicine, Power research group, and many others whom I have entirely forgotten (sorry).

I spent a lot of time thinking about neuroscience and rest at Hubbub, based at the Wellcome Collection in London (where I was supported by the Wellcome Trust). Writing and thinking with Felicity Callard, Hubbub's director, continues to be vital for how I think about the complexities of collaboration, dialog, and entanglement along disciplinary lines; so much of what I have to say on these topics (I gesture at this throughout the book) has either been learned from Felicity or remains inseparable from shared work with her. At Hubbub, I was lucky to find myself in the company of such brilliant people as Antonia Barnett-McIntosh, Josh Berson, Charles Fernyhough (and all the core group), Lynne Friedli, Nina Garthwaite, Harriet Martin, Ayesha Nathoo, Holly Pester, Charlotte Sowerby, Kimberley Staines, James "Seamus" Wilkes, and many, many (many!) more (given the group's size, I can only mention a few).

I am grateful to the many neuroscientists and psychologists—too many to mention individually, but they know who they are—who have put up with me in various capacities over the years. I learned a lot from sharing an ethics advisory board with Ilina Singh, Richard Ashcroft, Virginia Bovell, Raffaele Rodogno, and Sandy Starr. On the social study of autism, I have served on panels and coedited scholarship with several emerging stars of social research, especially Greg Hollin (a critical early reader of crucial parts of this book), Brendan Hart, Martine Lappé, and Dan Navon. I owe an immense debt of gratitude to Elizabeth Wilson and Lisa Blackman, who examined the PhD version of this text and who were gracious enough not only to pass it but to provide a range of excellent suggestions for its revision. At the University of Washington Press, I am grateful to Phillip Thurtle and Robert Mitchell, the editors of the In Vivo series, and to Larin McLaughlin, the press's editor in chief. None of the foregoing stops me from bowing to the convention (even if I am at least a little dubious of it) that errors and omissions remain my own.

Years ago, my friend Amy Hinterberger pointed out to me that it was odd behavior to write a manuscript about emotional labor without acknowledging the people in my own life who generate and sustain the affective structures through which things like book-writing take place. Nothing of any note in my existence happens without Neasa Terry, and this book is no exception. I am embarrassed to say that I only very lately realized that it never occurred to me to do anything *with* that existence other than write books like this one, and that perhaps my parents, Anne and Tom Fitzgerald, deserve some credit (or blame) for that. In either event, this book is dedicated to them.

Small sections from the setup to chapter 1 as well as a good amount of the substantive material in chapter 2 appear in "The Trouble with Brain Imaging: Hope, Uncertainty, and Ambivalence in the Neuroscience of Autism," *BioSocieties*, September 2014 (doi: 10.1057/biosoc.2014.15), published by Palgrave Macmillan. Chapter 3 is an edited version of "The Affective Labour of Autism Neuroscience: Entangling Emotions, Thoughts, and Feelings in a Scientific Research Practice," *Subjectivity*, July 2013 (doi: 10.1057/sub.2013.5), also published by Palgrave Macmillan.

TRACING AUTISM

INTRODUCTION

Looking for the Monolith

I N THE SUMMER OF 2015 THE NATIONAL INSTITUTES OF HEALTH (NIH) announced that it was joining a group of private and nonprofit partners to fund a $28 million research project looking for biomarkers of autism. Researchers based at Yale University were seeking objectively measurable biological signs that would, among other things, allow for a diagnosis of autism that was no longer based only on an assessment of behavior. The impetus behind the project was fairly clear: in recent years—whether as a biological, social, cultural, or scientific phenomenon—autism has become increasingly prominent. Yet the biological bases of its core features remain surprisingly elusive. In fact, despite many years of research, we still have no firm, coherent marker of autism, at either the neurobiological or genetic levels. This is a problem for researchers looking for autism interventions. It's especially a problem when researchers try to figure out what effect a potential intervention *has*. How, for example, do you measure the effect of an intervention across a group of people when—without a firm mark—it's not always clear that each person actually starts from the same biological place?

The NIH project is working on this question by using electroencephalogram (EEG), a brain measure, to track changes in autistic people over time. The "ultimate goal," according to the project leader, Yale psychologist James McPartland, "is to produce a set of measures that can be used as biomarkers of social and communicative function in ASD [autism spectrum disorders] and that could serve as indicators of long-term clinical outcome in clinical and drug development studies" (Peart 2015). Or, as he put it more succinctly in the *Hartford Courant*, the project is looking for "ways of quantifying human behavior that are not subjective and don't involve human clinical judgment" (Weintraub 2015). How is it that, in late 2015 by that report, our scientific accounts of autism are so ineluctably rooted to the realm of the subjective? How is that a developmental diagnosis can be so present and so visible, and yet also so heterogeneous and so mysterious? This is a central conundrum of neuroscientific research on autism. Sitting underneath that conundrum is a fairly widespread desire—however likely or unlikely individual researchers think it is—to identify *some* kind of stable, organic, and objectively measurable sign. For many, it is the neurosciences, especially neuroimaging, that will likely provide that sign: the research literature is replete with teams using one or more brain-imaging methods[1] to identify something in how the brain works, or how it is formed, or what it looks like structurally, or how its regions get connected that will offer—finally!—a definite and reliable sign of autism.[2]

In this book I am interested in that attempt. I am especially interested in what happens, and what it means, when researchers take a diagnosis as complex and as heterogeneous as autism and try to locate it more or less solidly in some function or substrate of the human brain. It is one thing to lament that autism is still identified in a surprisingly subjective, or at least nonbiological, way. But what would it actually mean to do research that would move the diagnosis onto some firmer terrain? How would that happen? What would it look like? What would actually be at stake in the attempt? Centrally at stake, of course, are biological and clinical questions about the identification of autism as well as questions of intervention (which remains controversial, for reasons I explore below). But it also seems to me that at the heart of this tangle of biological certainty and diagnostic ambiguity, there are important questions for medical sociology and anthropology and for science and technology studies too. Because what's also going on here is the production and working through of a series of tensions that intersect some of the most complex and important questions at the heart of those disciplines—tensions between autism and neuroscience, between biological markers and complex conditions, between bodies and lives, between objective measures and subjective accounts. Throughout this book I think about those tensions and the complex tangle that holds them together. What follows is an attempt to talk to and with a series of neuroscientists about how they think and feel their way *through* that tangle—and about the work that they do (conceptual, experimental, emotional) to pry it gently apart.

These questions emerge at what feels like a critical point in more than seventy years of scientific investigation. Autism was first described in 1943 by the child psychiatrist Leo Kanner, when he noticed increasing numbers of children coming to his clinic with a symptomatic profile that was not then well characterized within the psychiatric literature (Kanner 1968 [1943]). Often bright, with intellectual strengths in specific areas, these children (in Kanner's description) did not communicate typically, if at all; they often showed limited interest in the company of their parents or in anyone else's company; they used repetitive and stereotyped movements and often had specific interests; they had deeply idiosyncratic speech patterns and sometimes seemed happiest when left alone; so on and on, went the first account (ibid.). Building his description around a characteristic sense of *aloneness,* and on the degree to which in his view "these children had never engaged with the social world" (Grinker 2007: 53), Kanner characterized his syndrome as an "autistic disorder of affective contact"—borrowing the word "autistic" from a phenomenon that the psychiatrist Eugen Bleuler (1951 [1913]: 399) had observed in his adult schizophrenia patients: an "active turning away from the external world."

Kanner's clinic was not the only place that such patients made their presence felt. At more or less the same time that he was working in Baltimore—and apparently without either having knowledge of the other (or so was thought until very recently)—Hans Asperger, an Austrian medical doctor, observed and described a strikingly similar syndrome at his own clinic in Vienna.[3] Although Asperger placed more emphasis than Kanner on the strengths of his patients and described seemingly less severe idiosyncrasies, the clinical picture overlapped Kanner's with remarkable fidelity (see Silberman 2015; Wing 1981). This correspondence was broadly unknown in in the English-speaking world until Lorna Wing (1981) redescribed and reintroduced what became known as Asperger's syndrome. It subsequently emerged as an independent diagnosis, often popularly associated with "high-functioning" variants of autism (Asperger's has lately been folded into the more expansive category of autism spectrum disorder).[4]

As a category and a diagnosis, autism emerged simultaneously with the high point of American psychoanalysis (Luhrmann 2001; Hobson 2011), and although there was much debate in the early years about the degree to which autism was a "neurological" condition, its genesis was for a long time associated with parental—particularly, maternal—influences (see, e.g., Kanner 1949; Bruno Bettelheim's 1967 account remains the infamous landmark).

One of the effects of this assumption was to push the broader autism community, including networks of parents and allies, toward the creation and sustenance of their own parallel world of biologically oriented research and support. This eventually emerged as the clinical and research mainstream (Eyal and Hart 2010; see especially Silverman 2011). In the 1980s, as the influence of psychoanalysis waned, cognitive theories of autism started to emerge—for example, that the major cognitive phenomenon in autism is "weak central coherence," leading autistic people to focus on detail at the expense of piecing together the "whole picture" (Frith and Happé 1994; Happé 1996).[5] In recent decades attention has turned to biology, especially to neuroscientific and genetic methods. Brain-imaging studies, for example, have focused on brain areas, systems, or connections, where either function or structure might be reliably correlated with autism.[6]

Across these increasingly high-tech endeavors, Kanner's "autistic disorder of affective contact" remains recognizably the syndrome that today we know as autism spectrum disorder(s), the autism spectrum, or simply autism.[7] The most recent iteration of the APA's diagnostic manual, *DSM-5*, defines what it now calls autism spectrum disorder (incorporating the previously separate Asperger's disorder) across two major axes: "persistent deficits in social communication and social interaction across multiple contexts" (including things like "abnormal social approach and failure of normal back-and-forth conversation" and "abnormalities in eye contact and body language"); and "restricted, repetitive patterns of behavior, interests, or activities" (here are things like "stereotyped or repetitive motor movements," "insistence on sameness, inflexible adherence to routines," "unusual interest in sensory aspects of the environment," and so on) (APA 2013: 53). Autism is usually diagnosed in childhood, but it is a lifelong condition with a sometimes wavering course. It is strongly heritable, has some known neurogenetic components, is diagnosed in many more boys and men than girls and women (although it is still not clear why), and it is notoriously heterogeneous (that is, people at one end of an autism spectrum may lead high-achieving, fairly typical-looking lives while others require lifelong support).

In the decades immediately after Kanner's and Asperger's first diagnoses, autism research, unattached to any obvious neurological or other biological cause, remained a beguiling if niche interest (see Evans 2017). What is perhaps most remarkable about autism, though, is that in many countries, including the United States and the United Kingdom—growing slowly but really starting in the 1980s and 1990s—the diagnosis became very suddenly visible and very suddenly *present*. In 1976 autism had an estimated prevalence rate of between four and five per ten thousand children under age fifteen, or about one in two thousand or twenty-five hundred (Wing et al. 1976). By 2014 the Centers for Disease Control gave a prevalence of one in sixty-eight American children at eight years old (and one in forty-two boys) (CDC 2014). What caused this increase? Some have argued for an environmental factor, but the most common scientific view is that the increase in diagnosis reflects some combination of better recognition of autistic symptoms and "diagnostic substitution" of autism for older categories such as (in the American case) "mental retardation" (Frith 2003; Eyal et al. 2011).

Explaining this increase in cases of autism (the "epidemic") has been at the center of much sociological work in this space (see, e.g., King and Bearman 2011). But there are other questions that we have barely begun to consider. There is scope, for example, for more work on the material and cultural phenomena that emerged and became visible along with this surge. I am thinking of the quite sudden growth in popular awareness and concern about autism, across a range of cultural, political, and scientific spheres (see Osteen 2007). And there is a need for more attention to the things that did *not* become more visible, that did *not* become more present, even as autism occupied more and more of the public sphere. I am thinking about the absence of a clear technoscientific account of what autism is, what causes it, what biological mechanisms characterize it, how it might be treated, and so on. As Stuart Murray (2008: 2) has pointed out, as much as autism has emerged as a "pressing issue of current concern," so it also remains a "somewhat abstracted, unsourced, alien phenomenon." Which is to say: despite the increasing visibility of autism, and in particular despite concerted efforts to strategize research activity around it (Pellicano, Dinsmore, and Charman 2014), there is still no known single, identified biological cause, marker, or descriptor of autism's core symptoms.

Among other things, this means that diagnosis is typically done using a behavioral assessment only. This in turn raises questions about what exactly it *is* that's being diagnosed in such growing numbers, and even whether "autism" describes a biologically coherent phenomenon in the first place

(Geschwind and Levitt 2007; Happé and Ronald 2008). This has sometimes made autism research frustrating for scientists at the coalface: "the field of autism," noted the psychologist Laura Schreibman (2005: 7), "is littered with the debris of dead ends, crushed hopes, ineffective treatments and false starts . . . we are dealing with a devastating disorder for which we have few answers to date." More than a decade later, this does not seem like an inaccurate account of the research landscape. It remains striking that even as autism has emerged as a focus of popular concern, and even as it is located within both the bodies and habits of an ever-larger number of people, it has continued to resist any sort of easy clinical or biological definition. This absence is visible across a range of medical and scientific sites associated with autism. Jennifer Singh (2016: 87, 91), for example, has noted that despite a consensus in the 1980s and 1990s that "finding the gene or genes for autism would be relatively straightforward," today there is a palpable "sense of failure" in the autism genetics literature. That sense of deficit is felt, with special keenness, by neuroscientific researchers. In a review, Catherine Lord and Rebecca M. Jones (2012: 504) have noted, if there has been a sense of "great promise" surrounding neurobiological research, that research "has yet had little direct bearing on our understanding of what ASD is or how to treat it."

For the past few years, I have been thinking about the relationship between autism and neuroscience. Focusing in particular on the lack of biological certainty, and the anxieties that surround it, I have tried to get some purchase on how autism researchers think and feel about their work, as they move in, around, and out of research on the developing brain. This means trying to understand what it means—intellectually, methodologically, affectively—to take hold of something that seems as varied, as idiosyncratic, and as heterogeneous as the autism spectrum, and to try to bind it within the biological, statistical, and material constraints of the new brain sciences. It seemed to me, when I first started to think about this, that two outcomes were possible from this work: one is that autism would turn out to be a discrete, bounded, universal thing—a "natural kind," if we want to use that language[8]—that had only been waiting to be made visible in some neurological structure or function. The other possibility was that autism would turn out to be something much more (again, the terminology is barely adequate) "social"—a loose and contingent explanatory framework that we drape across particular kinds of people, even if only as a bureaucratic or taxonomic shorthand. My prediction (of course) was that it would be the latter. To be totally honest, and as a resolutely *social* social scientist, worried about biological reductionism, neurological imperialism, and so forth, I was assuming the latter. Indeed, to perhaps be unwisely honest, I was hoping that autism would *show neuroscience up in some way.* My assumption was that what my coauthors and I elsewhere have called the "epistemic murk" of the autism spectrum would not only prove quite elusive to neuroscientific methods, but actually in its very elusiveness would underwrite a critique of their narrowly biological and reductive presumption (Eyal et al. 2014: 236).

But what if this was not only the wrong prediction but also the wrong question? What if neuroscience was not a set of tools for revealing new truths about human beings, but neither was it a kind of organized, crudely biological misunderstanding of the endless complexity of human social life? What if the emergence (or otherwise) of distinctive pathophysiological patterns in the brain was not going to be adjudicated according to these terms? What if there was a very different story to tell about neuroscience and its relationship to a diagnosis like autism—a story that the more you inhabited it, the more it seemed to require suspending, and perhaps even giving up on, precisely the binaries that make such questions thinkable?

This is a book about scientists talking about their own practice, in tones that are beset by ambiguity, uncertainty, complexity, and even some anxiety. It has traveled far from the account of neurobiological imperialism that I set out to write. Of course there is nothing particularly shocking about this vision of neuroscience. Who today still clings to the fantasy of some thin, vitiated, dispassionate figure of scientific reason, neurobiological or otherwise? Yet there remains a question of what a different discursive register—complex, nuanced, tentative, sometimes contradictory—might actually mean in terms of how we think about the life sciences in the twenty-first century. More than a decade ago now, in her extraordinary short monograph *Psychosomatic: Feminism and the Neurological Body*, Elizabeth Wilson (2004: 27–29) proposed an image of the new brain sciences that is "different from those we have become used to in feminist, antipsychiatric, and social constructionist literatures"—one in which we can indeed begin to see an "articulate, obligated, libidinized" neuroscience as both ally and "resource" for social and cultural theory. Seeking this ally is a task that I want to take forward throughout this book. This is the story of a cutting-edge neuroscientific practice that is attentive to, and mindful of, the differences and limitations that run across its own practice. It is the story of an intellectual, practical, and affective complexity that has come to define the work of autism neuroscientists. It is the story of the care, seriousness, and novelty with which neuroscientists pick their way through that complexity.

TRACING AUTISM

Of course, interest in the growing relationship between the sciences of the brain and the diagnostic procedures of psychology and psychiatry is not new in the social study of the life sciences.[9] And yet surprisingly little light has been shed so far on what makes it possible, in the practice and epistemology of the neurosciences, to locate, mobilize, and follow complex diagnostic entities. I am referring to the way that neuroscientists have to engage in hard, ongoing work—experimental, conceptual, affective, and otherwise—on definitions, theories, methods, categories, and so on, to create convincing and sustainable accounts of the brain basis of complex diagnoses. I think there is still something worth finding out about what actually happens in that space, vis-à-vis the ways that convincing neurobiological accounts of often tricky, variable categories are actually made and sustained. We now know, to put it very broadly, that psychology and psychiatry are entering a neurological age (Andreasen 2001; Rose and Abi-Rached 2013); that this involves relocating more and more psychiatric and psychological diagnoses (and other styles of thought perceived as deficient) to the level of "brain disease" (Vrecko 2010); and that this process may actually recast how we think about, and intervene upon, categories of mental distress (Trimble 1996; Pickersgill 2011a). But still we do not have a full account of the ways that researchers in the new brain sciences actually think about, reason through, and *hold together* neurological accounts of complex diagnostic entities.

There are many valid ways to think about the new brain sciences. But at the heart of this book is a claim that this level, hitherto neglected, suggests something that we do not yet know about the putting-together of neurobiological accounts of mental states. As much as those accounts can be seen as processes of careful and accretive neurobiologization and neuromedicalization (Ehrenberg 2011), as much as they can be characterized as the techno-somatic whittling away of the complexity of human life (Pickersgill 2011a), as much as they can be theorized as an emerging and increasingly powerful neuro-reductive "vortex" sucking in all possible alternatives (Martin 2000), nonetheless, when you talk to neuroscientists working on the neurobiology of a disorder, you sometimes get a sense of the practice of those neuroscientists as something much more complex, much more ambiguous, much less monolithic, and much less certain, than the social science literature yet fully realizes.

The image that guides me is what I call "tracing autism." I first heard this term used by a midcareer psychiatrist and brain imager, whom I met toward the end of this project. He tried (without success) to get me to understand the difficulty but also the sense of possibility that surrounds image-based neurobiological work on something like autism. He described how a person might have a very "pure," identifiable and innate genetic lesion that disturbed their language functioning. Because of the person's language problems, the psychiatrist pointed out, people in their environment would react differently to them, so this very small and innate molecular difference would radically alter that person's social surroundings. This environmental input might then lead to a measurable biological difference elsewhere in their brain, which someone like this psychiatrist might then measure, but now as a relation to these other levels, and one that looped back into them in its turn. Such a structure of looping interactions between biological marks and the social world, somehow distinct *from* one another but also irretrievably entangled *in* one another, was what I had been trying and failing to understand. For this psychiatrist these looping entanglements were not a cause for despair, but they did lead him to a particular way of thinking about his work. "You can," he said, "without doubt, trace it up—now, not necessarily very easily. But you can."

I was struck by his use of the metaphor "trace." As a verb, to trace something is to doggedly follow its contours and turns. To trace autism, in this sense, would be to pursue it through its neurogenetic and environmental manifestations, to try to figure it out precisely through the turns that that pursuit takes. This sense of trace might be a near synonym for what another literature would describe as "constructing" autism. But what's interesting about "trace" is that it is a noun too. It describes not only the labor of following but the thing that gets followed: the trace, in this sense, is the half-seen original, the moving target (to bowdlerize a term from Ian Hacking [2006a]; cf. Navon and Eyal 2016). It is what guides us as we follow, the outline that edges the pen in specific, sometimes half-understood directions. What this borrowed image of "tracing" allows me to do—and quite unlike a metaphor such as "making"—is to describe the active way that these neuroscientists work to establish lines of connection within a complex and often ambiguous research area, while being faithful to how they *refuse* to relinquish the sense of a distinctive and singular neurobiology of autism as an organic phenomenon that is actually quite independent of this labor. I don't want to overdo this. There is no grand theoretical project here, no laborious conceptual rubric for others to follow. All I want this term to do is to remind me, through the conversations that follow, of the complexity with which these autism neuroscientists, looking for brain-based biomarkers, parse their relationship to (as it so often seemed to me) the intense ephemerality of their object.

What would happen if a sociology of scientific entanglement, contradiction, and difference was an argument for coherence rather than an entry point for critique? What if uncertainty didn't mean an entity was in question but actually signaled something closer to the opposite? It's not my preferred language—but what if being awkwardly, and sometimes confusedly, traced together, by a team of uncertain and emotional scientists, was what a "natural kind" actually *was*? In what follows, I want to think through some of the fundamental ambiguities, uncertainties, and differences that surround the work of neurobiological autism research, whether this is ambiguity about the scientific nature of psychology, a difference between the biological and diagnostic essences of autism, or an uncertainty about the efficacy of brain-imaging technology. I will try to be faithful to the complexity of the work involved in each, while recognizing that the neuroscientists in question are usually quite aware of this quality and of how they are moving through it. There are already important texts that talk about this sense of biological complexity and its ramifications—I think of Margaret Lock's (2013) work on Alzheimer's disease or Jennifer Singh's (2016) related work on autism genomics. What interests *me* is how researchers go on to think, talk, and work through these problems. How does this complexity come to *matter* for them, experimentally, ontologically, and politically? Leaning on the image of "tracing" helps me to show that what makes these scientists so interesting, and so worth thinking with, is that they do not take this complexity (or their own work of making sense of it) as an indication that one cannot talk about a broadly singular neurobiological disorder called autism. Perhaps, in fact, quite the opposite.

AFTER REDUCTIONISM

I first started to think seriously about these issues around 2008 and 2009. I remember beginning the project that forms this book with a kind of loose, jejune—and, needless to say, poorly read—antipsychiatric position. This was rooted in a desire to expose some of the reductive tendencies that I saw in the increasingly biological (and specifically neurobiological) thought style taking shape within and around psychiatric research. I was especially interested in how "the social"—at least as I understood it, as a kind of metaphysical space of interaction, signification, and experience—would become trapped by that thought style. I was concerned about what would happen as a series of psychiatric and psychological problems that were rooted in daily living, that were inseparable from the cultural and historical moments in which they were embodied, understood, and experienced, got more and more positioned as the universal, transhistorical, and material artifacts of an individual brain. If I was authentically naïve about all of this, the concern wasn't totally unwarranted: this was an era, recall, when serious books like *Brave New Brain*, by the neuropsychiatrist Nancy Andreasen (2001: x), assured readers that they might be "comforted by the fact that our social perceptions about [mental illnesses] have finally emerged from the Dark Ages . . . the scientific study of mental illness is now occurring in the era of the genome and the golden age of neuroscience."

Such views were, of course, subject to critique from scholars within the fields of anthropology, sociology, and history of science—most presciently and potently from the anthropologist Emily Martin. In an article published in 2000, Martin drew attention to how mind and body (nature and nurture) were, as she saw it, getting reconfigured in a range of new methods and concepts that clustered around the neurobiological and cognitive sciences. In the wake of such developments, she (2000: 576) warned, "the dyke between nature and culture has been breached, and all of what anthropologists call *culture* has drained through the hole."[10] What remains is a sense of subjectivity centered on a "universal" body—"unhistorical, unconscious of its own production, and possessed of many of the characteristics of modernist scientific accounts" (ibid.). With cultural and social explanation of human life all but ruled out by definition, "the brain becomes sovereign" and thus "generative of everything humans do" (ibid.: 574–77). What are we to do here? For Martin, context is key. She reminds us that the claim to an ahistorical body is *itself* a historically specific development, that the desire for universality is already highly contingent. If we want to understand neuroscience, we first need to understand "the environment we live in (and that scientific theories are produced in it) [which] had shifted so that a brain-centred view of a person began to make cultural sense" (Martin 2004: 200). Understanding neuroreduction means understanding "social life today" as it manifests in, for example, "the limitless profit and risk-taking of entrepreneurial, speculative capitalism" (Martin 2000: 578).

Emily Martin was—and is—not alone in this stance. In his important book on the rise of brain imaging via PET (positron emission tomography), Joseph Dumit (2004: 7) argued that much of the epistemological force of PET comes from its capacity to generate facts about persons that readily fit a preexisting mold—that is, a collectively held sense of an "objective-self" rooted in some of our "taken-for-granted notions, theories and tendencies regarding human bodies, brains and kinds considered as objective, referential, extrinsic, and objects of science and medicine." As Dumit (ibid.: 9) has pointed out: "What we come to receive as facts about ourselves are analyzable from a number of perspectives. We might look at the cultural salience of categories like mental illness and gender. We might look at the fundability of different approaches to brain-scanning. We might attend to the available metaphors for thinking about brains and people . . . it is out of this busy intersection of technical, social and cultural flows that scientists attempt to stabilize and conduct their experiments, and it is back into the intersection that their results must go." As Kelly Joyce (2008: 150, my emphasis) has put it—perhaps more strongly—in her ethnography of magnetic resonance imaging (MRI): "Imaging technology can be understood *only* by looking at the economic, social and cultural contexts that shape its meanings and uses." Manifestations of such context work are not hard to find. In their recent account of neuroscience and family policy, for example, the sociologist Val

Gillies and her colleagues Rosalind Edwards and Nicola Horsley (2016: 225) argue that "neoliberal reconfigurations of the state and society . . . can be traced right through into the narrow and anaemic biological conceptions of the social [that get] operationalized within the laboratory."

Perhaps so. But let me signal one issue that occurs to me here, which is that a critical interpretation of neuroscientific reductionism is sometimes met with an assertion on behalf of culture, or history, which seems, on the face of it, no less reductive or simplistic in its orientation to the world. In other words, for every injudicious neuroscientific claim about our capacity (one day! soon!) to trace every petty human phenomenon to some structural or functional component of an individual's brain, we will find an anthropological, sociological, or historical claim that is no less ontologically simple— that every petty human phenomenon, including all of those associated with the brain sciences, are always (and always have been!) only the working out of ineluctably social and cultural patterns. Perhaps I am being unfair here (Joseph Dumit, for one, explicitly stresses that such is not his view). Yet, as Felicity Callard and I (2015) have argued elsewhere, there is often a striking symmetry between neuroenthusiasm and neurohorror, as each form of attention tries to pinion the other's interests within its own object of concern—a kind of intellectual thumb war in which the outcome is unclear, but the stakes are agreed: one of neuroscience and culture must come under the thumb of the other.

In the introduction to her 2014 book *The Autistic Brain* (coauthored with Richard Panek), the well-known autistic scholar and writer Temple Grandin relates the story of her childhood diagnosis and her sense of gratitude that her mother had taken Grandin to a neurologist rather than to (what may well have then been) a psychoanalytically inclined psychologist (Grandin and Panek 2014: 8). For Grandin, understanding what's going on with a person with autism means understanding what's going on with their brain: "the problem wasn't a psychic injury," she writes, "and Mother knew it" (ibid.: 9). Grandin is highly optimistic about the "hard science" of neuroimaging and what it will reveal about autism, which, as far as she is concerned is "in your brain" (ibid.: 20). Grandin does not hold back from spelling out the implications of this view. Recalling some of her early encounters with MRI imaging, she describes the sense of excitement she got from looking at the images that were produced: "Seven or eight times now," she says, "I have emerged from a brain-imaging device and looked at the inner workings that make me *me*: the folds, and lobes, and pathways that determine my thinking, my whole way of seeing the world" (ibid.: 22).

Grandin reminds us that autism tends to disrupt our usual ways of thinking about the political, social, and affective sequelae of neuroscientific discourse; when we think about autism, we start to understand some of the ways that a brain can get figured, not only as a winnowed space of narrow and fatalistic reduction but sometimes, too, as the site of a lively cosmopolitanism, of an intensely engaged and "quirky" citizenship (Bumiller 2008). What emerges around autism (but not only around autism) is a sometimes fraught neurological politics of identity. If this politics is neither straightforward nor uncontested (see especially Orsini 2012), nonetheless it should bring us up short when leaping to easy conclusions about what it is that neuroscience is going to *do*, politically and culturally.

I began this project with a desire to contribute to the literature on neuroreduction. My working hypothesis was that the definition of autism as a disorder rooted in specific patterns of social interaction and communication (among many other things) had created a category that from the outset, and however well constructed the assessment tools, would *inevitably* be so intensely heterogeneous and varied, so freighted with symbolism and meaning, so marked by culture and history, that it would not—could not—be corralled into the narrowly organic requirements of the brain sciences. It became more and more apparent that the neuroscience I encountered was very different to the one I had gone looking for. Where I expected simplification, I found complexity; where I expected certainty, I found only ambiguity; where I expected arrogance, I found self-effacement and awkwardness; where I expected optimism and expectation, I found anxiety about the future; where I expected unbending scientism, I found affectively weighted ways of understanding, narrating, and thinking through scientific practice. Most important: where I expected a kind of committedly reductive organic materialism, I found a group of neuroscientists who were intensely aware (far more than I ever was) of the capacious thing they were dealing with, and who were profoundly open to figuring out a mode of scientific practice and experimentation that would remain alive to that complexity. What I found in the practice of autism neuroscientists was not a misguided attempt to put some very different categories of thing together; what I found was a series of intellectual, practical, and somatic strategies—and not all of these necessarily admirable—for producing, living with, working through, and even locating some scraps of scientific positivity in the complex and entangled ambiguities that run across neurobiological autism research.

REPARATIVE NETWORKS

Is it is possible that we have obscured something important in social scientific accounts of neuroscience? Is there any risk—to paraphrase Eve Kosofsky Sedgwick (2003: 124)—that how we have learned to talk about the brain has made it less rather than more possible to understand the political, epistemological, and ontological potentials that inhabit it? Sedgwick was of course working in a very different milieu, interested in unpicking (as she saw it) the forms of "paranoid" reading that often animated critical (and especially queer) literary theories—in lieu of which Sedgwick proposed a "reparative" practice that might work through some more pleasurable and ameliorative registers (ibid.: 144). Sedgwick's essay, much read and much contested, has produced a rich seam of literature within queer and feminist studies on what precisely is entailed by the call to reparation and the affects that glue it together (see, e.g., Love 2010; Wilson 2015). I will not rehearse those debates here. Instead, and at the risk of some simplicity, I want to dwell on Sedgwick's original formulation because I think there is still much to be gained from thinking about what the call for a reparative, over a paranoid, style might yet do for the social study of the life sciences. Reparative reading, as Heather Love (2010: 237) has glossed it, directs us to "multiplicity, surprise, rich divergence, consolation, creativity, and love." Against grandiloquent fault-finding, the reparative urge "stays local, gives up on hypervigilance for attentiveness; instead of powerful reductions, it prefers acts of noticing, being affected, taking joy, and making whole" (ibid.: 237–38).[11]

Will this help? Will it do? On the very day I sat down to write this paragraph, a new edited volume, *Neuroscience and Critique: Exploring the Limits of the Neurological Turn* (De Vos and Pluth 2016), arrived in my pigeonhole for review, with an endorsement from no less a figure than Slavoj Žižek himself ("a book for those who still dare to think!"). "Who is taking up the critical task today with respect to the neurosciences," the editors ask in their introduction, "and what kind of critique of them is possible?" (De Vos and Pluth 2016: 2). To which all one can say is, Who indeed? Jan De Vos and Ed Pluth are no simple reactionaries; they do not seek to only repel neuroscientific influence on the humanities, and their self-conscious notion of critique is rooted much more in the spirit of a careful mapping than a vulgar conceptual battering. Yet still, even in 2016, it seems we cannot escape this long and deep desire for critique of the neurosciences; if we have managed to cast off some of the vulgarities of previous years, it sometime seems like we have replaced these with a quest only for more and more sophisticated forms of paranoia.

Ways of reading neuroscience that I am tempted to call "reparative" (none of these claim this term for themselves) have started to emerge: notably, Nikolas Rose and Joelle Abi-Rached (2013: 233) have argued that on the other side of criticism is "opportunity"—that "the human sciences have nothing to fear from the argument that much of what makes us human occurs beneath the level of consciousness or from the endeavours of the new brain sciences to explore and describe these processes." The anthropologist Nicholas Langlitz (2010: 40), in a study of neuroscientific research on hallucinogenic drugs, has described his own realization that "the prevalent objectivist image of cognitive neuroscience had to be qualified to apply to this particular case—and possibly not only to this case." I have more to say about all of this throughout this book, but let me briefly gesture here at some of the ways of thinking about science that have more broadly guided my inquiry. This is a set of writings and practices that sometimes goes under the label of a "material feminism" (e.g., Alaimo and Hekman 2008).

As I suggested earlier: Leaning on an image of "tracing" is going to help me show that what makes these scientists so interesting, and so worth thinking with, is that they do not take this complexity, or their own work of making sense of it, as an indication that one cannot talk about a singular neurobiological thing called autism. There is clearly strong influence from the strange quality that Karen Barad (2007: 33) has called "intra-action." For Barad, talking about "intra-action" helps us avoid a commonsensical assumption about independent things *interacting* with one another and describes, instead, a world in which "distinct agencies do not precede, but rather emerge through, their interaction" (ibid.). Just because things are complex and ambiguous, and sometimes only held up by their entanglement with other things, it does not follow that one has to deny them their agency, or their distinction. Just because the neurobiological account of autism seems inseparable from the interactions of (and not the distinction between) biological truths and diagnostic convenience, or just because an account of autism's brain basis only comes *after* the difficult weaving of psychological concepts into physical science, this does not require us to concede that there is no such thing as a neurological, brain-based autism, distinct in itself. A related argument has been made by Victoria Pitts-Taylor (2016: 35) in her account of brain plasticity—a phenomenon that allows us, she says, to stop figuring the brain as "either nature or culture; rather [the plastic brain] is a specific configuration of matter and meaning that achieves itself in entanglement with the world." As it is for Pitts-Taylor, Karen Barad's conceptual apparatus, which I discuss in more detail in chapter 2, is critical for helping me to understand how a neuroscientific account gets made, without reducing it to the status of the made-up.

But perhaps the single greatest influence on what follows in this book comes from the work of feminist theorist Elizabeth Wilson (1998, 2004, 2015). What is at stake for Wilson (2004: 14) in her long and deep theoretical engagement with psychology, psychoanalysis, neuroscience, and feminist theory, is the possibility that "feminism can be deeply and happily complicit with biological explanation . . . that feminist accounts of the body could be more affectionately involved with neurobiological data." She points out that in the interplay between neurological, psychological, and feminist theory, the first part of that tangle does not always have to play the "catastrophically doctrinaire" role ascribed to it (ibid.: 28). This is a strikingly different vision than that provided by the sociological commonplace that cuts biology from its "constitutive relations with other ontological systems," thus rendering it "isolated and destitute" (ibid.: 70). If such a vision sometimes seems distressingly omnipresent in the social sciences and humanities, Wilson (ibid.: 70) points out, it is not limitlessly powerful: "The barriers behind which biology has been sequestered do not annul the secondary relations that biology has on other systems (e.g., the effect of neurotransmitter uptake on psychological mood), and it is these kinds of causal relations that neo-Darwinian commentaries seek to exploit. These barriers do, however, obstruct the operations of a more originary relational network . . . within which biology is constituted, animated and evolved."

I am in pursuit of this network here. I am equally in pursuit of an "articulate, obligated, libidinized" neuroscience, which "may even be a resource for theoretical endeavour, rather than the dangerous and inert substance against which criticism launches itself" (ibid.: 29). In her most recent book, *Gut Feminism*, Wilson (2015: 1) extends this project to ask: "What conceptual innovations would be possible if feminist theory wasn't so instinctively anti-biological?" A related inquiry structures this book: What conceptual innovations would be possible if the social study of the life sciences wasn't quite so determinedly *social*?

INTERVIEWING NEUROSCIENTISTS

What follows is the product of thirty-seven semi-structured interviews that I carried out between 2009 and 2010, mostly with UK-based autism neuroscientists but with some people in related fields too.[12] I say "neuroscientists," but this is maybe a less obvious a category than it first seems. As Steven Rose (2004: 3) has pointed out, the coming-together of a series of specialisms and fields—cognitive psychology, neurology, neuroanatomy, and so on—into an agglomeration that today we call "neuroscience," is both a recent and an unstable phenomenon. It's true that I interviewed at least one graduate student whose program carried the title "neuroscience," but she represented a fairly recent phenomenon. Most of the others had advanced degrees in psychology; some had medical degrees and worked as psychiatric researchers; at least a couple (people primarily interested in neuroimaging methods) had PhDs in physics; some practiced clinically as well as doing research, but many didn't. All of which is simply to say that how and whether one gets called a "neuroscientist" (by me or anyone else) is less the sign of a clear intellectual division than it is a function of interest, bureaucracy, and time. In chapter 4, I discuss the memoir of a well-known neuroscientist who charts the evolution of his job title from "psychologist" to "cognitive neuroscientist." That chapter is where I most fully dig into these overlapping affiliations and sites of attention, and where I consider their history. The reader interested in this story, or who is seeking the traditionally historical first chapter (which, otherwise, I do not supply) might be well advised to start there (although I don't recommend it).

There are two things worth stressing at the outset: one is that that I take a fairly ecumenical view of "neuroscience" throughout this book. Everyone I spoke to (1) had a major interest in autism (sometimes it was their *only* interest); and (2) used neuroscientific methods (and often *only* neuroscientific methods) to pursue that interest. In all cases, interviewees whom I call neuroscientists either did research in a neuroscience center, frequently contributed to neuroscientific projects, or had a significant publication history of neurobiological studies. Some may find this strategy inadequate, but the truth is that if you want to get some hold of "neuroscience," today, you have to remain fairly open about who is authorized to represent that practice. The other thing that's worth noting is that, despite such openness, when I talk about "neuroscience" here, I mostly mean "cognitive neuroscience"—that is, the branch of neuroscience most closely identified with brain-imaging methods and that draws much of its intellectual inheritance from psychology.

Absent (or mostly absent) from this book are the more overtly biological branches of neuroscience (cellular and molecular neuroscience, for example) as well as other subfields, such as computational neuroscience or neuroanatomy. I don't want to make too much of this. My initial trawl for neuroscientists working on autism in the United Kingdom overwhelmingly threw up people working more or less in the tradition of the cognitive neurosciences—a bias in the field that, again, becomes pertinent in chapter 4. More importantly, as Rose and Abi-Rached (2013) have pointed out, to rigidly parse the distinctions between the neurodisciplines, or the practices contributing to those disciplines, might be to miss the point. What matters is the practice and style of thought suspended in that common prefix *neuro*. So it will be remain for me here: I am interested in scientific researchers who affiliate to this style of thought and the material practices that go with it, irrespective of their job title or their disciplinary background.

A couple of other things worth knowing about the interviews that underpin the book: two-thirds of the interviewees were women, a fact that I think is inseparable from the tilt toward cognitive neuroscience and psychology, which I say more about in chapter 5. These interviewees were spread across a wide range of age and experience: I interviewed two graduate students, twelve researchers at the postdoctoral or research-fellow level, nine at the level of lecturer (this would be the equivalent of an assistant professor in the North American system), and eleven at the level of reader, professor, or principal investigator (associate or full professors in the North American system). A typical interview lasted around forty to forty-five minutes. In general, I proceeded inductively, without a set question list but with a set of core themes (these changed over time), including such questions as: Why was it important to "find" disorder in the brain? What were the barriers or problems to this? How do biology and psychology relate to one another? Why was there so little consistency in the autism neuroscience literature? How did the interviewee get involved in the field? Where is the field going? And so on. I did not rely on a strong distinction (either temporal or conceptual) between the work of conducting interviews, the analysis of the interviews, and the formation and reformation of my research questions.[13] As I progressed, nonetheless, this procedure produced a spiral of overlapping investigations (compromising, essentially, accounts of why it was important to do neuroscientific work on autism, what that research was like, and what the interviewee's own attitude to it was). I transcribed and coded the interviews, established major themes and subthemes, then went back and forth recoding and rearranging those as necessary. I did this until I was left with five major questions that seemed to more or less fit the data. These comprise the five substantive chapters of the book.

Before getting into that substance, let me say something about interviewing—and especially about interviewing as a method orthogonal to the strong tendency toward ethnographic (or, at least, ethnograph-*ish*) methods in social studies of science and medicine. This book is in many ways an instance of a now somewhat unfashionable genre—the "interview study."[13] Of course I visited people in their labs and hung out to whatever minor extent I was allowed (generally none). I went to talks and chatted informally to people. I kept up with the literature, scholarly and otherwise. Like any normal person, I lurked—I continue to lurk—on Twitter. But it is the interview—the talk—that remains core to my account of what's happening in this space. I know that I risk looking defensive here (obviously I *am* defensive), but I am aware that, for some people, this will be a problem. Without a rich ethnographic attention to the day-to-day practice of the laboratory, without the spatial and temporal richness of fieldwork, without an upclose micro-attention to how people actually talk and act with one another *away* from the performative nonspace of the interview room, how can I presume to speak, in any meaningful way, about what's actually going on? Such concerns are not new. As Mike Savage (2008) has pointed out, the interview, long tainted by its association with psychology and social work (being the favored instrument of psychoanalysts, health visitors, and other traveling moralists), has had a central role in the history of jurisdiction in the social sciences. This suspicion rubs against another uneasy history: the development of sociological method, and the interview method in particular, as an instrument for governing particular kinds of populations through the elicitation of sentiment (see Rose 1999: 70).

Social scientists interested in neuroscience have tended to come from two other angles instead. One is the broad and "public" world of neuroscientific discourse: the public statements of neuroscientists, the broad discursive claims that structure the field, and the arenas in which neurobiological assumptions get read into public policy, education, and similar spaces (see, e.g., Gillies, Edwards, and Horsley 2016). The other comes via ethnographic encounters with neuroscience: with the laboratories, conferences, workshops, and meetings that make up its day-to-day practice (Beaulieu 2000; Roepstorff 2004; Cohn 2008). Both of these angles are deeply important, of course. The first is ideal for following, assessing, historicizing, and qualifying the emergence of a discursive field of "neuroscientific truth" (see Rose 1996b). The second has shown us how the neuroscientific everyday belies the public image of the field, including perceptions of it held by neuroscientists themselves. But how should we think about neuroscientific concepts and practices that are not yet "in the true" and yet that are also known, and talked about, and mulled over, on the surface of everyday discourse? I think especially of how people, in a scientific field, substantively think about, reflect on, talk around, and reason *through* (or not), the sticking points, confusions, disappointments, ambiguities, inconsistencies, oddities, potencies, and so on, of the material-semiotic practice in which they find themselves.

There's nothing original or profound about this. But it's important to remember that something happens when people tell you things. There is, as Ann Oakley (2016) has pointed out, something of "the gift" about the moment of the interview. It's hard to describe in advance of the material in the main chapters, but it seems to me that there is something about the interaction that surrounds the interview—the formalization, the presence of the recording device, the sense of enclosure, the ticking clock, the list of questions, the back-and-forth . . . frankly the whole on-the-surface naked *performance* of the thing—that opens up a way of thinking about science and scientists that is *just very different* to the more usual ways of engaging these fields.

And not just thinking about but thinking *with*. In what follows, I am motivated a lot (often implicitly) by thoughts of collaboration, entanglement, and generally nonparanoid ways of talking, thinking, and working *with* other agencies.[15] Interviewing is an important part of that commitment. In many ways this is a problematic claim, but there is something to be said for the interview as a way of taking a practice, or a person, or an idea, *seriously*. In her commitment to the surface of talk, in her refusal to dig down into the subterranean realm of everyday practice—in her attempt, that is, to accept the gift of that "which insists on being looked *at* rather than what we must train ourselves to see *through*" (Best and Marcus 2009: 9, emphasis in original)—I find the interviewer to be a *much* more compelling figure of nonparanoid engagement than the participant observer.[16]

I don't want to overdo this. Obviously the distinctions I draw are not hard and fast. There are many dreadful—and paranoid—ways to conduct interviews (no doubt I have conducted several such myself). As is well known, ethnographers and others are perfectly alive to the nuances and problems of their practice, and they continue to sustain a lively dialogue around it. But still I want to stake a claim for the interview-with-a-scientist as an analytic object in its own right. Interview data should not be seen as supplementary to ethnographic entanglement but rather as something worth gathering, and analyzing, in its own right. What's at stake in the interview, including the work of coproducing and analyzing that interview, should not be at the mercy of ropey gestures about ethnographic presence, including the fantasy of *thereness* that such gestures rely upon. Suspending that fantasy, I focus on how a series of neuroscientific researchers, taken as a cohort more or less, and all of them working on autism, talk about the kind of work that they're engaged in. My field *is* their account of the conceptual and practical labor of doing a neuroscience of autism; it is made up of these scientists' thoughts on their successes and failures, their ongoing working out of the connections between autism and neurobiology, their still-in-progress ideas about what needs to be done for this connection to be established more firmly—and their speculations about how such labors might yet more widely recast our ideas about a diagnosis like autism, and a practice like neuroscience, in the first place.

I start, in chapter 1, with neuroscientists' accounts of what kind of thing autism actually *is*. When I pursued this question with them, I got two very different kinds of answers: On the one hand, autism was described as something that was "biologically true" and that was composed of an "unchanging core." On the other hand, they described autism as heterogeneous and separable, a "symptom check-list" or an "umbrella of convenience." But rather than the people I spoke to feeling the need to resolve this tension, I show how the links between a biological singularity and a dispersed intangibility were traced together by thinking about autism through alternative registers. Across this description, I show how "neurobiologization" is a more various, complex, and uncertain process than is sometimes allowed. I show how a broader process of "biomedicalization" may be made up of some strikingly recursive and oscillatory moves.

If chapter 1 is about neuroscientists talking about autism, chapter 2 is concerned with how they talk about neuroscience itself. My interviews show how this apparently hard and reductive science, often associated with a kind of neurobiological chauvinism, is also sometimes deeply inflected by feelings of uncertainty, disappointment, and even anxiety. I interpret this as a kind of contradictory and hesitant attitude in which research programs are somehow assembled around a deflationary sense of *low* expectation. I argue that a "tracing" neuroscience, one that proceeds through a strange entanglement of confidence and ambiguity, may well be bilingual in discourses of both promise and failure. Chapter 3 extends this discussion by looking at how the categories of "knowing" and "feeling" can get mixed up in the neuroscience of autism. Working through researchers' personal memories of the field, and the registers in which those memories are related, I show how people I spoke to sometimes gave accounts of their own intellectual biographies that mingled a commitment to good science with the affective and emotional aspects of scientific work. I show how they narrated their work through registers of upset and heartbreak, desire and excitement, sometimes even a "visceral" commitment to research.

Chapter 4 is about psychology. I focus on how some of the people I talked to described psychology as a discipline once embroiled in "Freud and faff," but one that is now, under the aegis of a physics-based cognitive neuroscience, unambiguously scientific. I approach this observation via Georges Canguilhem's idea of "recursive" history, and I suggest that these claims to science might be heard as forms of boundary working within a contemporary psychology, newly troubled by the reemerging salience of something like "social context." In chapter 5, I think more concretely about the mechanics of actually tracing things together within neurobiological autism research. I consider especially a series of metaphors used by my interviewees for their own neuroscientific practice, where they variously described it as a process of "tangling," "assembling," "connecting," or "shuffling." Drawing on the work of Bruno Latour, Karen Barad, and especially Donna Haraway, I argue that that these metaphors are ways of talking about a "contingent stability" in autism neuroscience, a practice that actually allows autism to be traced across precisely these different levels. Such contingent forms of stability describe *both* a key activity of autism neuroscience in its own right *and* an explanation for what lies beneath the sometimes fuzzy-looking commitment to difference and ambiguity that I have tried to describe.

There is a lot of talk in what follows. But this is a book about the lived complexity of neuroscientific practice, as it is talked about, talked through, remembered, explained, regretted, justified, and laughed about by neuroscientists themselves. My analytic object is not neuroscience behind the veil. I have no interest in yet another account of what really goes on behind the scenes at the laboratory. What I am in search of, instead, is what neuroscientists think they're up to, what the strange practice of neuroscientific experimentation looks and feels like to them, and how they talk about it when they are asked. This book is my account of what I heard, when I talked to autism neuroscientists about what it is they actually do.

1

THIS THING CALLED AUTISM

Between Biological Disorder and Diagnostic Convenience

O N THE FACE OF IT, ASKING A GROUP OF AUTISM NEUROSCIENTISTS, "What is autism to you, exactly?" is not the most amazing research tactic. After all, autism is a reasonably well-described and well-understood clinical phenomenon. When I started writing this book, I was fairly sure I knew that autism was a neurodevelopmental spectrum disorder with characteristic deficits in three core domains (the autism triad) of social interaction, communication, and repetitive behavior (APA 2000). I would have said fairly confidently, if you had asked me, that autism was strongly heritable (Bailey et al. 1995), that it was diagnosable in about one in eighty-eight school-age children in the United States (CDC 2009), that it was much more prevalent in boys and men than in girls and women (Baron-Cohen, Knickmeyer, and Belmonte 2005), and so on.

But none of this is as straightforward as it looks. For one thing, as noted in the introduction, changes to the *Diagnostic and Statistical Manual of Mental Disorders* (the *DSM*, a major clinical and research tool used to identify autism in many countries) have removed communication from the three domains entirely and have collapsed the distinction between autism and related diagnoses like Asperger's syndrome (APA 2013). The one in eighty-eight figure has had to be revised too—a more recent report increased this to one in sixty-eight in the United States (CDC 2014)—while there is also now a suggestion that the much-observed gender difference in autism might be, at least partly, a difference in gendered *presentations* (Werling and Geschwind 2013). More to the point, perhaps, there is ongoing disagreement about whether the so-called triad actually comprises a specific disorder at all (Happé, Roland, and Plomin 2006), while the "specific genetic etiology" of autism, after more than a decade of research, "remains largely unknown" (Gupta and State 2007;

cf. Singh 2016). Skating across all of this is a series of political contests over whether we might characterize autism as a "disorder" in the first place, and even if the diagnosis might not mark a "cognitive style" or a talent (Happé 1999). "We know more about autism now than at any point in history," notes cultural historian Stuart Murray (2011: 1), "yet, at the same time, if we're honest, the foundational observation that we might make, the 'central fact' about autism with which we should probably start, is that we don't know very much about it at all."

DIFFERENCES

As the sociologists Daniel Navon and Gil Eyal (2016) have pointed out, when we talk about autism, we are in the realm of what the philosopher Ian Hacking (2006a) calls a "moving target." In other words, we need to pay attention to the ways in which "the disability we call autism . . . change[s] its contours and its lived experience" (ibid.). It was while trying to get a handle on this terrain, and the movement across it, that I became interested in the kinds of things that neuroscientists would talk about, if you just asked them what autism *was*. I wanted to know, in particular, if they took autism to be a fairly discrete biological entity, whose roots and causes we might one day discover given the right tools, or if, instead, they took autism as a temporary and contingent array of symptoms, whose current classification likely floated free of any singular bodily substrate. I especially wanted to know if any of this *mattered* for autism neuroscience, or whether the definition of autism was only a kind of philosophical book-keeping. I wanted to know, to the extent that it *did* matter, how these researchers dealt with this question, how they rationalized it, felt about it, and so on. When I asked this, a lot of people said that autism was a kind of biological truth with an unchanging core; others thought it was just a checklist of symptoms or a diagnostic convenience; still others weren't so sure either way, but they were fairly certain that there was something distinctive about an encounter with an autistic person, and that this was maybe all the surety they needed. One of the overarching goals of this book is to think through some of the ambiguities, uncertainties, and differences that are thrown up by neuroscientific research on autism. It is well known, now, that the clinical objects of both the "psy" and the "neuro" sciences are rather less stable, and rather more strangely inflected, than the public cultures of those disciplines will yet fully admit (Dumit 1999; Blackman 2007).

But what I want to do is not only show a fairly large definitional ambiguity at the heart of autism neuroscience, although such is certainly the case. I also want to begin thinking about the ways in which neuroscientists actually work *across* such ambiguities. In particular, I want to show how neurobiological research on autism might progress, not despite definitional and diagnostic differences but somehow with them and even through them. This is at the heart of what I describe as "tracing autism"—the act of pursuing, enacting, and enabling a firm neuroscience of autism precisely *through* forms of difference, ambiguity, and entanglement.[1] "Tracing autism" means not being put off by these kinds of collective ambiguity. It describes a conviction that you don't have to carefully pull things apart if you want to establish some kind of scientific singularity or separateness. You don't need to distinguish so carefully between biology, idiosyncratic behavior, diagnostic checklists, and clinical pragmatism; you don't have to so parsimoniously distinguish the brain and its disorders from the bureaucratic exigencies of the clinic.

Within the social study of psychological and psychiatric diagnosis there has been a good deal of attention to *neurobiologization,* or the process of locating some otherwise fuzzy categories of human and social life firmly within the bounds of an individual skull (Abi-Rached 2008; Williams, Katz, and Martin 2012). This process is real enough. The sociologist of parenthood Pam Lowe and her colleagues (Lowe, Lee, and Macvarish 2015), for example, have convincingly described how health and welfare policies in the United Kingdom have become enraptured by a crudely reductive, neurobiological account of child development. But how does neurobiologization actually happen, from the point of view of neuroscience? Is the process of making a neurobiological account of something as crushingly inevitable, and as crudely reductive, as we sometimes think? In this chapter, I shift attention away from an analysis of increased certainty, toward something we know rather less about: the set of vagaries, ambiguities, complexities, and sheer idiosyncrasies that are nonetheless *intensely* visible at the heart of these projects in biomedical inevitability.

I also extend this complication of the discourse around neurobiologization into a larger argument about *biomedicalization*—a term that describes the emergence of a widespread, quasi-industrial, preemptive space of ethical intervention within contemporary biomedicine (Rose 2001; Clarke et al. 2003). It strikes me that if the emergence of autism as a neurobiological category says something about the potential biomedicalization of one tricky diagnosis, it also shows this process to be much more complicated than is often imagined. It shows how biomedicalization and neurobiologization, far from being unidirectional processes of increased certainty, actually take place via some remarkably recursive and oscillatory movements. It shows how, even in the twenty-first century, a neuroscientific account can maintain a promiscuous relationship to the tightening loops of biology and technology.

THE UNCHANGING CORE

Early on in this project, I would quite directly, and very naïvely, ask researchers whether or not there was a real sense of "biological truth" to autism. By "biological truth," I think I meant an autism that was not only a concatenation of clinical symptoms with a more or less variable neurogenetic core, but an autism that was in some way "biologically discrete." I meant an autism, in other words, that would ultimately be located in a malfunction in, or idiosyncratic pattern of, some specific brain circuit or function—such as the (then popular) mirror neuron system (Dapretto et al. 2006)—or else in a series of specific genetic mutations, such as the 22q11.2 deletion from chromosome 22 (Fine et al. 2005). Basically I was interested in the idea that if you might once have answered the question, "What is autism?" by saying that it was a product of overeducated parents (Kanner 1968 [1943]), or a reaction to maternal rejection (Bettelheim 1967), or a symptom of childhood schizophrenia (Rutter 1968), or an outcome of "mindblindness" (Baron-Cohen 1995)—in the mid- to late 2000s you would more likely talk about something like the ratio of white to gray matter in the brain (Herbert et al. 2004), or the thickness of the corpus callosum (Just et al. 2007), or something similar. This is the view that, as one scientist I spoke to put it, autism "has to be a brain thing, in that, if you could fix that thing in the brain, if you knew what it was, then you could, I guess, cure it."

"The question is, of course," said a senior professor to me, doubtless growing tired with my attempts to dance around this topic, "if there's a kind of natural entity that's autism." "Yes," I conceded. "Who knows?" she said eventually, but went on:

> I think that there is. But I sometimes doubt it. . . . I have that view [that autism is a natural kind]. Why do I have that view? I think it dates back probably to the sixties, when I was no doubt educated to see it like that, and it has in a sense stood me in good stead. And it was possible even to have that narrow view that we had at first about autism to expand it, and to embrace a whole spectrum of conditions. That's pretty good, you know—it sort of seems a strong, robust sort of concept.

The committed subscription to a model of autism-as-natural-entity is striking, particularly given how aware this researcher is of the relationship between her view and a particular historical way of thinking about things, as well as the way in which this view is for her basically pragmatic. When I asked this professor whether there had not been a lot of changes in how we thought about autism over time, she readily agreed that there had been "huge changes," but then said: "it is a wonder that we can still recognize the same thing, but I think we can, which is remarkable." This "same thing" of autism is what I was trying to get at with these kinds of questions. And indeed I frequently found that people were committed to this view of autism as a kind of "thing"—singular, persistent, biologically discrete.

Where people are committed to this biological truth view of autism, they often point to its persistence through history. Talking about her early work in the field, another senior person said to me: "People whom we saw then, what, forty years ago, we still have contact with now and they still meet criteria," Or as another put it:

> Historically you can go back and find cases of people who clearly had autism two hundred years ago before anyone knew the concept of autism—so, as an organic, developmental brain disorder, it exists in the same sense that schizophrenia exists, cerebral palsy exists, Down's syndrome exists, you know, and has always existed.

In his part-memoir/part-history *Not Even Wrong* (2006), author and autism-parent Paul Collins splits the story of his own autistic son's diagnosis, with the story of Peter, the nineteenth-century "Wild Boy" of Berkhamsted, through whom Collins draws parallels with the autistic children of today, even enough to diagnose, in the book's subtitle, a "lost history" of the disorder. Elsewhere, the social historian Rab Houston and the psychologist Uta Frith (2000) tell the story of Hugh Blair of Borgue, a "natural fool" of eighteenth-century Scotland, in whom they are inclined to make a tentative retrospective diagnosis. Indeed, the alignment of their project with arguments about the essential and time-less nature of autism is remarkably explicit: "We believe that it is important to separate the existence of labels and explanatory theories from the existence of pathological conditions. The syndrome now called autism was not categorized until the 1940s. . . . however that does not mean that [this condition] did not exist before then—just as there were presumably germs around before germ theory was promulgated to explain disease. . . . through studying Hugh Blair's condition in its historical context we hope to be able to expose *the unchanging core of autism*" (ibid.: 4; my emphasis).

This "unchanging core" is precisely what I mean when I talk about the biological truth of autism. Certainly, none of the scientists whom I spoke to expressed this view easily or flippantly. But there were plenty for whom autism had some kind of organic and timeless singularity at its heart: "If you break it down enough," a postdoc said to me, "you would find biological markers in common. . . . I doubt it would be at this whole-umbrella level. It would be at the subtype level. But I'm sure there are . . . there is a biological truth. But I think it . . . yeah, I think it's difficult." What I find interesting about this statement—and I don't mean this at all in a critical way—is how *unconvincing* it is, even to itself. "There is a biological truth," she says, almost like an invocation—a faithful quality that precisely explains the concession that comes almost immediately after: "it's difficult." The commitment to bio-logical truth is indeed just that—a commitment. And, as we shall see, not always one that is easy to maintain.

Indeed, the sense of how unconvincing this view is, even to its own adherents, is among the most remarkable things about it. I had come to this research at least in part prepared to tell a story about biomedicalization—an analytical framework established by the sociologist and STS scholar

Adele Clarke and her colleagues (2003). Biomedicalization builds on *medicalization*, a now classic concept in sociology that describes the way in which a once-social phenomenon (often some kind of deviant behavior) is brought discursively within the purview of a medical science (typically psychology or psychiatry), thus creating the potential for a reconfiguration of forms of expertise, intervention, and government (see Conrad 1992; Lupton 1997). But with this view having become something of a sociological cliché (Hedgecoe 1998), Clarke and her colleagues propose a series of important additions. These include the emergence of radically novel medical technologies and new fields based on these technologies; the shift in medical focus to the maintenance of health over the cure of illness; a medical-institutional bureaucratic concern with prevention and risk assessment; the emergence of novel corporeal possibilities and social movements extending from these; and the situation of contemporary medicine within a hypercapitalist political and economic complex (Clarke et al. 2003: 62–63; see also Clarke et al. 2010).

This is a powerful frame, and it does indeed capture much of what is happening between neuroscience and autism. But it nonetheless seems to predict (rhetorically if not directly) an ever-tighter intertwining of the fuzzy, heterogeneous category that we now understand as autism and these emergent biomedical and political-economic agencies—precisely on the basis of autism's deepening subscription to new "bio-" technologies and fields. As the sociologist of psychiatry Jackie Orr (2010: 379) has put it for the field of US psychiatry more generally, biomedicalization describes "an uncanny generalization of the techno-structures of a 1960s mental hospital into a widening assemblage of medical and social spaces." This kind of generalized bioessentialism—"it has to be a brain thing"—was certainly prevalent in my conversations with autism researchers. I met people content with the emerging biological truth of autism and also people quite content to generalize the "techno-structures" of neuropsychiatry into this space. But these accounts were heavily diluted by a sense of autism as something that often manages to *evade* new technologies in the biosciences and as something that *resists* the widening field of biomedicalized psychiatric and psychological generalization. In such accounts, autism is *not* obviously in the process of becoming a more stringently biomedical concern. Sometimes, indeed, quite the opposite.

IT'S NOT CLEAR THAT THERE'S JUST ONE CONDITION

One of the most striking things for an outsider entering the field of autism research is simply how much is missing from our knowledge of autism—knowledge, for example, about things like cause, course, the degree of heritability, the likelihood of effective therapy, and so on. But perhaps even more remarkable is the extent and proliferation of explanatory gaps between the things we *do* know. In other words, not only do we not really know what causes autism, or how it manifests in the brain of an individual, or how it affects cognition, or even what it will look like in any given diagnosed person. It is also unclear how the few things that we might know, at any of these levels, will connect or interact with what we know at any *other* level. You and I may manifest similar idiosyncrasies in communication and interaction, but this is no predictor of a shared cognitive or neurological problem in the end.

This, in simplified form, is one element of what scholars usually lament as the "heterogeneity" of autism (Ronald et al. 2006). But autism is heterogeneous in a number of distinct and important ways. Most strikingly, autism is heterogeneous in terms of its symptom set: the famous (if now superseded) diagnostic triad of communication, language, and repetition can appear to very different degrees within equally diagnosed people (Ring et al. 2008). There is also a recognized sense of heterogeneity in how well a diagnosed person will "function" in what some call the "neurotypical" world (i.e., the world of "typically-developing" people). Many people diagnosed with an autism spectrum disorder lead perfectly independent, typical-looking lives; others will never speak and require permanent care (Howlin et al. 2004). And there is heterogeneity of causal pathway—autism is highly heritable, for example, but specific mutations for more than a minority of diagnoses continue to elude researchers (Betancur 2011).

Thus it is very possible, and indeed likely, that people who share broad symptomology in early adolescence—enough to diagnose each with autism—may have arrived there by very different developmental routes. All of this is common knowledge, and scholars have tried to deal with it in a number of ways. The clinicians, and specialists in neurodevelopment, Mary Coleman and Christopher Gillberg (2011), for instance, have started a trend for referring to "the autisms" or "the autistic syndromes" (cf. Geschwind

2009). Indeed, such is the degree of disparity between the presence of the core symptoms, some have suggested that there will likely be *no* overlap in genetic causes between the different components (however big the sample size), and the best outcome might be to abandon the idea of a single "autism" altogether (Happé, Ronald, and Plomin 2006).

None of this is terribly controversial or astonishing to people in the field: "No one thinks of a unitary thing called autism," said one notably blunt psychiatrist to me when I first raised this question: "You've got a crazy mind if you think there's a unitary thing called autism. . . . It's a diagnostic category just like any other diagnostic category. And you can draw the line on that category where you wish to draw the line." This is, for many, the rather public way of talking about autistic heterogeneity—and also the way that it is often reconciled with the "biological truth" view, discussed earlier in this chapter. But I also often found people who were less sanguine. This was because not only was research failing to reduce the causal and explanatory gaps between the different elements of autism, but sometimes it seemed to drive those gaps further apart: "The more the science tells us," a former autism charity employee told me, "the more complicated the picture gets, and the less likely we are to find simplistic solutions." As a senior professor put it:

> My current beliefs about autism, my intellectual beliefs about autism, based on the data we have, is that the underlying basis of the social and communication difficulties is going to be different from the underlying basis of the . . . repetitive behaviors and the special interests and to some extent the detail focus . . . if there ever comes a time when we can intervene, we'll be able to intervene separately on those different components.

A more junior person put it even more starkly:

> I think that there won't be one single cause for autism, or for any other disorder probably. That might mean that there is a constellation of biological markers that, if you're born with those, you're likely to develop autism under certain conditions—and those conditions may well be completely unmeasurable and out of our grasp. I don't think they'll be as straightforward as, I don't know, you were exposed to x substance at x point in your development or anything. I think it'll be a very complex interaction of

things. And that complex interaction will be different for many, if not all, individuals. So I think this search for a single cause at whatever level you're looking—whether you're looking at a cognitive level or a more biological level—I can't see how that is possible. I think if it was possible, we would be much clearer as to what it is. And I don't think we would see so much heterogeneity.

Far from research bringing us closer to autism's biological truth, it is increasingly clear that what looks like a unitary thing called autism may actually be the result of multiple-level interactions between genes, environments, and behaviors—and these are very unlikely to have a single biological identity. "Certainly a few decades ago," said one lecturer, "people didn't think it was a biological condition at all. They thought it was a response to a particular parenting style, or who knows what else. But now there's very good evidence that there is a genetic basis." But then she said:

> Although it's actually a complicated genetic basis—so one problem is that although it seems to have a genetic basis, it's not a single genetic basis, it's not clear that there's just one condition. We don't have a clear marker the way we might do for sickle cell disease, we don't have a clear genetic way of defining the condition in terms of identifying particular genes. So it may be one condition that varies in intensity. Or it may be a variety of conditions that share a common genetic load.

I am particularly interested in this contribution because the interviewee begins with a biological truth position, but then immediately, and once she starts to talk in detail about the biology, starts to qualify this commitment and instead emphasizes the knowledge that in fact, according to her, "we don't have." This is precisely the kind of oscillation that I am concerned with in this chapter—that is, the shift between a fairly straightforward idea of biological truth and a more problematic and complex view of autistic "heterogeneity." I stress, I am not saying that the latter is by definition problematic for neurobiological research. What I am trying to draw out, nonetheless, are these strange, recursive, and sometimes anxious relationships between biology and multiplicity. I am interested in the growing recognition that there may in fact be no "there there" in the biology of autism.

And yet "an autism" persists. When I asked one scientist, a fairly junior lecturer, what a final "neuroscience of autism" would actually look like, she said:

> It's not going to be a point change in the brain . . .it's going to be the influence of a small change here on a small change here. Individually, you won't be able to observe those changes; they'll be too small. But their interaction will be observable. And in fact it is observable—it's called autism [*laughs*].

I love this joke—or at least I love the observation that of course we can already "see" the thing that's apparently so hard to find. How might we account for the stubborn persistence of sameness ("it's called autism") through this discussion of endless heterogeneity? In a short primer aimed at an interested lay public, Uta Frith (2008: 4), one of the most senior and well-known autism researchers in the UK, recollected: "When I first saw autistic children I was only dimly aware that autism comes in degrees, from mild to severe. Actually all the cases I saw were severe . . . [now] autism is no longer a narrow category but has widened enormously to embrace a whole range of autistic conditions."

But then, after reflecting on the implications of this realization, she goes on: "Every individual is unique in a multitude of ways, but they also resemble each other in some fundamental preferences and characteristics . . . no one has yet given up the idea that there is a common pattern behind the kaleidoscope of individual behaviours. I will therefore frequently use the familiar terms autism and autistic, as a reminder that there is a central idea behind the spectrum" (ibid.: 4–5). Here we see the same move: enormous width on the one hand; fundamental characteristics on the other. When I spoke to a very eminent professor about these issues, she said, referring to her general philosophical orientation, I think, and only half seriously:

> Probably . . . I'm a bit naïve . . . I think that we are each of us individuals in the sense that we are incredibly different from each other, through our genes, our history, all sorts of factors that shape us—the physical environment. And, yet, you can forget about all this and say, "Well, here is a man [*laughs a bit*] or a woman." And even though your hands and my hands are very, very different, I'm sure, they have five fingers [*I interject to point out that*

I have notably small hands—which is true]. Never mind—you know, there is a hand . . . so I suppose philosophers would call that essentialism. Anyway, I am sort of quite, you know, I see that all the time in my own looking—I'm not particularly worried about the differences. I can see the differences and I think they are what makes life interesting and everything, very nice. But I am not detained by them to such an extent. And that applies even to autism.

I am really struck, upon rereading this, by her ability to scientifically reconcile questions of difference with some kind of identity nonetheless. I very much like this interviewee's response to my own—genuine—qualms about the smallness of my own hands, namely the straightforward, taxonomic reassurance: "Never mind . . . there is a hand."

What I'm trying to show here is that when a group of neuroscientists talk about autism, they sometimes talk about the degree to which it is a somewhat unitary and distinctive disorder, recognizable even in history, and with some sort of basic, biological underpinning. But entangled with this biological view is a sense of autism as a much more heterogeneous and separable phenomenon. This latter autism has a biological pathway so radically unclear that it seems likely any biological truth will be found only within a series of quite dispersed and contingent events—events that are perhaps even distinct from one another—that we only diagnose as autism when they (maybe arbitrarily) co-occur. But I'm also trying to show the way in which somehow, despite this, "an autism" is held together, distinctly, through the two registers. Below, I have much more to say about the way that an autism neuroscience traces its way across these two registers. But before moving on, I want to push the discussion of these two narratives a little bit and delve a bit deeper into the consequences for the scientific practices that are formed and reformed around them.

THE ONLY TRUTH THAT PEOPLE AGREE ON

Let me, for a moment, belabor one lengthy discussion, which is taken from an interview that I did with a young postdoctoral autism researcher who had trained in psychology in another country but who was now working in a psychiatry department in the United Kingdom—a series of facts to which she steadfastly refused to attribute any epistemological significance what-soever. It was a bit of an odd interview, conducted in a psychiatric research institute, and in the course of what became, for me, a long and melancholy afternoon spent hanging around the grounds (I was supposed to interview one of this person's colleagues at the end of the day too). I was about to write that it was one of my longest interviews but, having checked, I see that it was actually a notably brief encounter—just over thirty-five minutes.

I remember this scientist as being a small bit combative and challenging, and, like many of the people I interviewed, somehow torn between suspi-cion of my motives and contempt for my inanities ("Oh my God," she says at one point on the tape, entirely exasperated, "I mean, if I would be able to answer these questions . . ."). But it was also weird because by the end of our conversation she sounds (on the tape and also in my memory) quite exhausted, even a bit despondent about where we had ended up. It was as if this was the first time (and I say this with no sense of self-aggrandizement; indeed, I take this as evidence only of my blundering) that she had talked out loud about some of the trickier aspects of her own practice. I relate the conversation here in some depth because the critical parts center on pre-cisely the questions I have outlined: questions about what exactly autism is but also, more important, about how a neurobiological account of autism gets traced through some very different registers.

We had started off talking about different ways to think about and approach autism, and she had defined her own method as natural-scientific, specifically neuroscientific—saying that she was interested in the "potential to answer questions in a not so fluffy way." We danced around this for a bit ("fluffy"?), and after a while I said: "Can you give me your most useful definition, off the top of your head, of autism—that you work with? Not the textbook definition but the definition that's practically useful to you." She responded:

Well, I would always give, and I usually do in talks, the definition that
is given in the diagnostic process—which is that autism is diagnosed
based on a triad of symptoms based on social interaction communication
and restricted interests. And I'm using that because it's very convenient.
Nobody's questioning it because if you really start thinking about it, then it's
very difficult—*really* defining autism. Because it is not only diagnostically

defined as a spectrum, for example, of abilities—low-functioning people and high-functioning people, they are all within the spectrum—but also that between individual differences are so large that I find it sometimes quite difficult to put all those people into one umbrella term. And so far it's a very convenient way of defining it by just going back to the *DSM-IV.*

"Is that how you think of it?" I asked. "Just a kind of umbrella of convenience?" She said:

Yeah, of course—I think many colleagues are also questioning . . . well, I think there is a certain agreement, but there is not one autism, yeah. There are certain things which are shared between people on the spectrum, but there is definitely not one sort of autism.

This gap between "a certain agreement" and "definitely not one sort of autism" closely mirrors what I talked about already. Of course, there is a historical context to this conversation—a context in which (though it goes unmentioned in the interview) accounts of autism have in fact shifted over time; this sentiment of multiplicity also seems inseparable from the fact that this researcher and I were talking right around the transition point between two diagnostic manuals, which would shift the definition of autism not insignificantly. But I am more interested in another gap that opens up here. This is the gap between this researcher's earlier commitment to an uncomplicated science that quantifies to reduce *fluffiness* and the fact that her actual research object might be a bit of an arbitrarily cut-off assemblage of different phenomena; and anyway, no one could even agree whether it existed or not, beyond some "certain agreement" that nonetheless remained undefined. "So," I said, happily mixing metaphors, "the umbrella is really a flag of convenience for research purposes, for diagnostic purposes—but really, we think underneath this, there's a series of different things, or . . ." I let the question peter out. She replied:

There's a series of different disorders. Different—and, again, it's very difficult to differentiate them because basically that's what I'm trying to criticize in a differentiation between, let's say, autism spectrum disorders and other developmental disorders. It's of course an artificial differentiation. So it

would be equally wrong to say "autism spectrum disorders are made up of
five or, let's say, six different disorders"—because then you just reintroduce
an artificial differentiation. I think there are spectra of abilities and dimen-
sional domains and you tick some boxes and some you don't as an individual, and if you are on the spectrum or not, even that, if you are looking at
the whole population, if you are diagnosed with autism or not, is sometimes
a matter of degree.

"Okay," I said, "that's interesting. I guess . . ." Here, I am tentatively trying to
come to the point. "I mean something that's interesting about that is that it
sounds kind of *fluffy* to me."

It does, it does.

"So, I'm intrigued," I said.

It's not that I like it.

"Okay, um . . . so how do you deal with that then?"

I think what you do—what like, constantly, that's the whole purpose of
being a neuroscientist I think, is to try to overcome the fluffiness and just
try to formulate something that can be regarded as temporary truth, and
you work with [that].

"But it seems to me that as the neuroscience of autism progresses . . . there's
less certainty. That it's . . . more answers become possible rather than fewer."

Yes, that's quite difficult to cope with, it's true. Yes, like, there are other
theories that it's a theory of mind issue, theories like weak central coherence and so on, and they all have their niche in autism spectrum disorders . . . but there is nothing so far which really [*inaudible*] the potential to
cover it all. Yeah, and we go further away from it.

Here, a tension opens up between, on the one hand, "the whole purpose
of being a neuroscientist" and, on the other, the fact that as research

progresses "we go further away" from autism. "Yeah," she said again, when I pushed it.

> I think we are not working towards one final truth. I think that's too chal-
> lenging for us. We will never get there. Like neuroscience in general and
> with autism spectrum disorders, let's say, well what would that truth—there
> are *sometimes* final truths like with, um, let's say Down's syndrome has a
> certain truth behind it. Well then again that's maybe, I don't know, maybe
> questionable—there are differences between individuals with Down's
> syndrome.

"I guess," I said. "What's interesting about Down's syndrome is that you can point to a particular chromosomal problem and say 'this is it.' You don't think we'll ever do that with autism?"

> There [are] many, many genes which are involved, it seems. There's no
> universality.

"Not just on the basis of genes, though." I said, "Do you think we could ever point to a brain area or a function or a something and say, 'This is autism?'" She responded:

> Well, I think the only truth that people agree on is that that's not the case.

I am really interested in this exchange, because not only does it give us this "no universality" view of autism in a very clear way, but this researcher clearly relates how the lack of "final truth" is both a property of "neuroscience in general" and something that's quite "difficult to cope with." In his discussion of personality disorder, the sociologist of psychiatry and neuroscience Martyn Pickersgill (2011b) has shown how neuroscientists sometimes manifest an ambivalence about their own efficacy, rooted in the intractability of this diagnosis. For him, this should open up a question about what it is that biology is doing, precisely, within the contemporary brain sciences. Something very similar is at stake in this conversation, in which the story of autism's putative neurobiologization is neither straightforwardly one of victory nor is it a lament for failure. It comes instead in the form of a rolling, hybrid account, in which the researcher's commitment to neuroscience sometimes runs up against the daily reality of being an autism researcher.

Later on in this interview, this scientist and I talked about gender and its relationship to the brain. Socialization, she explained, was a big issue for developmental neuroscience in general because it was impossible for the researcher to isolate where or when exactly the almost infinite constellation of possible interactions between genetic and environmental inputs were woven *just tightly enough* to form the kinds of culturally identifiable and interesting shapes that we might tentatively name "autism" or "male." "Yeah," I said, "so I mean—so it's a particularly difficult science, I guess, is what's interesting about it." There was a bit of a pause: "A very fluffy one," she eventually said.

Let me stress that it is not my purpose to enter a critique of developmental cognitive neuroscience as a particularly "fluffy" pursuit (which, at any rate, would be a shortsighted move for an interpretive sociologist). I *am* interested in what has happened in the space of a conversation that lasted only a little over half an hour and in the context of a discussion only about the definitions that are at stake in neuroscience research on autism. This researcher has moved from an uncomplicated self-narrative in which she becomes a neuroscientist precisely to distance herself from the "fluffy" behaviorists and psychoanalysts who had once dominated the discipline in her home country, to an account of developmental cognitive neuroscience, her chosen area, as now itself a decidedly complex if not compromised intellectual pursuit. How is it that that talking about "what autism is" to a neuroscientist brings such tensions to the fore? How do people deal with this uncertainty? How does a more or less stable and researchable autism persist across it?

IF YOU'VE GOT DOWN'S SYNDROME, YOU'RE NOT FASCINATING

Because what's really striking to me about all of this is that no one throws up their hands in despair. Even if no fact about it is uncontested, still my interviewees trace a coherent autism through both the categories of "biological truth" and "umbrella of convenience." To think about how this is done, I introduce another category that was consistently deployed in the course of discussions about what autism is. This is the category of enigma and, extending out from it, the idea of autism as an object of particular fascination. In fact, the term *enigma* is strikingly common in discussions of the neurobiology of autism.[2]

Someone once said to me, explaining how she had altered her research trajectory after obtaining her PhD, that she had "brainstormed on a piece of paper the different things that I was interested in, and out of that came that I had this real intellectual interest in autism that had always captured my imagination, the kind of 'the enigma' and all that." The "and all that" is telling—it suggests a shared understanding that the word is now even a bit of a cliché in autism research, and that just by using it, the researcher can gesture at and assume that I'll instantly understand the long chain of associations. *Enigma* itself is an interesting word because it doesn't just suggest mystery; it specifically implies a puzzle with a ludic element to it, something "to afford an exercise for the ingenuity of the reader or hearer" (Oxford English Dictionary 2016a). This quality was frequently expressed by scientists in terms of their "fascination" with autism. "There's a certain element to people with autism that is intrinsically fascinating," said one, "that sparks your interests and makes you ask: 'Why are they the way they are? What is going on inside them?'" Or as another said of her first encounter with autistic children in a special school: "It's just so different. . . . I found it completely captivating."

What is it about autism, specifically, that gives it this fascinating, enigmatic, captivating quality? One scientist said to me:

> I think the fact that autism is essentially about social engagement makes it a different disorder. As a society we . . . social contact is absolutely crucial, and we can't really comprehend that you wouldn't want to have that contact, for example, and our world doesn't work without that social interaction at varying levels. So I think there's something just quintessentially different about autism, that's to do with this social instinct, if you like, that we must have, that people with autism either don't really have at all, or have in a really unusual way.

Or as a graduate student put it:

> I think it's, it's really an interesting disorder, because it's kind of every-
> thing that's sort of dysfunctional in autism is kind of what makes us kind of
> human, if you like. . . . I think it can tell us a lot about, sort of, how we are
> as humans generally, as well as the actual autistic conditions.

Autism is fascinating—such is the suggestion—because "what autism is" is no
less than a privileged microcosm of human development in general. But what
interests me here (what *I* find fascinating) is how researchers figure autism as
a mysterious gateway to something much larger—and how this makes all of
the ineffability, the weirdness, and the complexity somehow much easier to
live with. In these descriptions the contradictions of autism are configured less
as barriers to knowledge and more as hints of the great secrets that lie behind.
An association with some enigmatic—maybe even an unknowable—human-
ness allows researchers to bridge these thorny questions of biological essence
and distributed heterogeneity. One senior professor said to me:

> The first time I met these [autistic] guys, I thought, "Wow," you know? They
> embody kind of all the things that puzzle me about psychology. Because it's
> a developmental disorder—in other words, there's something different about
> their constitution, which doesn't determine, but constrains the way they
> develop, so they develop differently, along a range of different trajectories,
> which have certain things in common. And so it touched all the buttons,
> I was just, you know, "What's it like to be you?" I mean, I've always been
> fascinated by that about, you know, anybody, any other person you meet:
> "What's it like to be you?" But these guys, being so different. They *are* so dif-
> ferent. And yet so similar, you know?

This sense of "so different and yet so similar" is a critical link for connecting
the diagnosis of autism to the question of the human. People with autism
are enough like neurotypicals for useful comparisons to be made; but they're
also different enough, and different in the right ways, for the distinctions to
hint at something larger and more interesting. By drawing attention to these
registers, I don't want to focus too much on the solidity of these links (or,
indeed, their politics, which are to say the least complex). I'm not at all interested in (in fact, I am troubled by) the cultural idea that autism is some kind
of chink in the veil of human uniqueness. But what I want to think through,

nonetheless, is how this sense of intrinsic and ongoing fascination might explain how researchers persevere with neurobiological studies of autism, even amid the inevitable difficulties, confusions, and setbacks.

Consider the following interview, which I pull out at some length, because it opens up this issue in a really explicit and interesting way. It comes from a professor of psychology who had trained at the Maudsley Hospital and (what is now) the Institute of Psychiatry, Psychology and Neuroscience (IoPPN; the Maudsley and IoPPN, in South London, form part of perhaps the United Kingdom's most prominent psychiatry and psychology research and teaching facility). "Yes," she said when I brought this topic up, "it is intrinsically fascinating." She went on:

> I mean, like these early children that my colleagues talked about at the Institute of Psychiatry, we would be talking about, over coffee, this child that might, you know—we were very young in these days; you wouldn't get on a course as young as it [today]. We were totally naïve, and in our early twenties, and didn't know anything about anything. And then there would be these children that would do really weird things. It was the mismatch between often the appearance of being quite bright and then not really having any common sense. But also occasionally you'd get children who had these, um, you know strange behaviors, repetitive behaviors or obsessions, or knew strange things. I remember this one child I was asked to assess with autism, sat there saying, "You don't want to do this." I sort of thought, "No, I don't." And then I thought, "No, actually, what he's . . ." [words swallowed by laughter. What she's talking about is the sudden and unexpected acuity of the child with autism]. And I thought it was great, you know, "Woah!!" So it is very . . . but again I think I'm partly driven by the fact that I'm such a neurobiologically oriented person. I wanted to know, you know, what in the brain generates some of these behaviors.

I remember hearing this last interjection—"I'm such a neurobiologically oriented person"—as the re-erection of a kind of scientific boundary against the rather human and personal turn the interview had taken. Which maybe should not be a surprise. The historian of psychiatry Anne Harrington (2005), in her work on neurologist Oliver Sacks, has written about the romance of the individual case history and how it can trouble a borderline between the humanistic and the strictly neurobiological: "the particular, the emotional, the value-laden, the meaningful, and the relational aspects of human experience," Harrington (ibid.) suggests, "functions to remind us that being a human being—a human brain—is still a more complex and richer thing than can be contained in the spare and reductionistic vocabulary and frameworks of our sciences."

This actually makes me think about my interviewee's coda as a *deliberate* mingling of the relational and the scientific. I hear it as a way of bringing some neuroscience back into this conversation about fascination: neurobiological research, she was emphasizing, was not outside of, or inimical to, or something that had to be disentangled from, her fascination with the enigmatic world of human difference. Indeed, it seems to me that, far from marking a boundary point, neuroscience emerges here as a space in which a person might think through the relays between her memories of science and fascination. It was the study of autism in particular that allowed these qualities to comingle. The same person said to me later on:

> I mean, this is the sad thing. If you've got Down's syndrome, you're not
> [scientifically] fascinating. . . . everything is just deficit. You're like a younger
> child. So, in autism, you get behaviors that are not just like a younger child,
> you get behaviors that are unexpected—and the other thing, when I first
> started to work on [this], and I still remain absolutely gripped by it, is that
> what they emphasize is what is weird, and often that is language, if we're
> not autistic . . . so, if you get these children who are very overliteral, so, um,
> there's some wonderful, wonderful examples—one being a teacher saying
> to a child at school just sort of "just go into the toilet and get yourself a glass
> of water," and the child is sort of getting the water out of the toilet bowl. But
> then you think, "Well, why don't we all do that?"

She went on:

> You feel that what the autistic child is doing is actually more logical than
> the rest of us. They're taking a literal interpretation of what you've said, and
> then it forces you to think, "How the hell do we operate and not do that?"
> And I regard that as still a huge question of enormous interest to me. If I
> could crack that I'd be very happy. And then there is restricted interests—

which are so odd, I mean, you know, why are . . . I mean, okay, some of
them are just like normal little kids only more so, but, you know, when you
get these kids that are fascinated by drainpipes, or lampposts, you know,
what the hell is that all about? It's just sort of . . . if you're not fascinated
[*laughs*] . . . so, you think that there's . . . there's something there that potentially, if you could understand it, [you'd] get to the bottom of what it's like to
be human.

What attracts me to these extracts is the way that this interviewee ties
together the themes of the intrinsic interest of autism, its admirable qualities, and the privileged insight into the human it grants. But at the heart of
this story is the tracing of a now memorialized intuition that, through these
different things, *there's something there*—and that this must be something of
"enormous interest" to her as a "neurobiological person."

I am not at all suggesting that people talk about "enigma" to strategically
avoid hard questions or uncomfortable truths. I am suggesting that creating
space for the enigmatic is actually what makes it possible for a neurobiologically oriented person to work through contradictions that would otherwise
be hard to stomach. It is what allows this scientist to trace firm neurobiological understandings across those differences. The enigma of autism, situated
in memories of clinical encounters and autistic idiosyncrasies, allows her to
trace some coherent sense of autism, and of its neurobiological description,
through time, past her own bewilderment, and across some significant personal and intellectual differences.

THERE'S THIS THING WE RECOGNIZE WHEN WE SEE IT

Several times, after people had talked (at some length) about the heterogeneity of autism, about its trickiness as a disorder, and about the lack of
certainty surrounding its existence as a clinical entity in the first place,
there was a bit of a pause. Then they would say, nonetheless, that there was
something distinct and knowable about autism all the same, even when this
commitment could only be articulated as a kind of *feeling* or as something
that you *just knew*. I first heard something like this from a child psychiatrist
who was also an active brain-imaging researcher. I actually had had quite

an awkward and hurried encounter with him, arranged at short notice. The conversation skated across the kinds of unremarkable, mainstream, public-facing accounts of autism neuroscience with which I had by then become bored. I never really felt as if I had broken through the professional veneer. However, I eventually asked him—because he maintained a clinical psychiatric practice in addition to his research interests—about the difference between diagnosing autism for research versus diagnosing it in the clinic.[3] He said:

> Yeah well, in the clinic, as you say, there are often people who fall short
> on one or more of [the standard autism scales used in research], and then
> you've got to use your clinical judgment to decide whether the level of
> impairment they have is sufficient to warrant the diagnosis despite falling
> short on one or more of those tools, or their, sort of, how they feel to you
> as a clinician, then the feel of somebody with Asperger's, despite falling
> short. . . . I suppose the difference there from research is that, to actually get
> a paper recognized by the community, the research community, they look to
> see that people have been positive on the criteria. Whereas if you're doing it
> in order to work out what's best for that person, it's a bit different.

I was quite taken with the use of the word *feel* in this context, especially from such a (I say this as one to another) apparently straitlaced person. So I asked him to elaborate a bit. He said (predictably):

> Yeah, *feel* is the wrong word because you don't actually use your hands to
> do it, because this is more about . . . [*Longish pause. I stupidly butt in: "Clinical
> skill?"*] Yeah, because with autism, it's a syndrome, so there's a collection of
> signs and symptoms. It's like a pattern almost. People will have a number of
> this very long list of signs and symptoms. And somebody who meets a lot of
> those, who has a lot of those symptoms, but not others, they'll have a cer-
> tain way that they come over—so they'll for instance be using some of the
> language, they'll have some of the language features, for instance, of autism.
> Or some of the social features of autism. And it's that mix of features which
> makes somebody *feel* to a clinician whether they do or don't have a disorder,
> in some ways.

There are a number of interesting issues here. One, of course, is that the autism of the clinic is not necessarily the autism of the clinical trial, which is a well-recognized (albeit unresolved) issue within randomized control trials (RCTs) of psychiatric diagnoses (Zimmerman, Mattia, and Posternak 2002). What counts as a diagnosis for research purposes, where homogeneity is key, might be different to what counts for an individual's diagnosis in a regular clinic, where what matters isn't so much the reliability of the description but actually figuring out how best to help that person. This is crucial. But perhaps more interesting, in *this* case, is what marks the difference: for inclusion in the clinical trial, participants have to pass a given cutoff on at least one, and preferably both, of the "gold standard" quantitative scales. This is a requirement of publishing in a good journal and is obviously governed by concerns about the homogeneity of participant populations across different studies. But for the clinic, where this kind of specificity is less of a concern, there is a different solution, which is to cede some epistemological space to whether autism is actually *felt* by the clinician in the course of the encounter. If I have been interested, up to now, in how autism remains something qualitatively distinct and knowable for these scientists, even when all research hitherto has failed to ultimately isolate it biologically, here I focus on the way that interrogating a *feeling* helps the researcher or the clinician to trace the neuroscience of autism across this tension.

Of course, these issues are not unique to autism. They track long-standing debates about the art and science of clinical diagnosis—and, in particular, discussions about "evidence-based medicine," worries about the dissolution of the gap between clinic and lab, the emergence of a "clinical science" to service this rupture, and so on (Gordon 1988). In one sense, this is the intellectual and political context in which we should read the willingness of some clinicians and researchers to talk about their recognition of autism in terms of its *feel*. But I am also keen not to simply re-create those long-standing discussions, which are well established in both the medical-sociological (Mol 2008) and medical-medical literatures (Malterud 2001). Instead, I wish to focus on this insistence on the qualitative distinction of autism as a biological entity, which allows these scientists to trace autism across the questions of "unchanging core" and "symptom checklist." I argue that this tracing is (at least partly) done via the partial deferral of "what autism is" to some sort of *feeling* of autistic presence. Thinking through the "writings, films and statements of those autistic individuals who seek to represent themselves," Stuart Murray (2008: 33) has insisted on both the multiplicity and indelibility of a specifically autistic "presence." This presence he argues, "extends beyond the ways the condition is labeled in medical and other institutional contexts" (ibid.). I am entirely with Murray in this insistence. But what I think is at stake in my encounter with this young clinician is also a quasi-*medical* commitment to something like that presence and a willingness to see it, and to feel it, even as it goes unmarked by a range of well-validated clinical measures.

Consider this exchange that I had with a junior neuroscientist—a postdoctoral researcher in a metropolitan psychology department that had a heavy research concentration on autism. He said to me, in the middle of a fairly frank discussion about autistic heterogeneity:

> There is something about each individual with an autism spectrum disorder that makes them part of the autistic spectrum. They do share certain difficulties, and certain areas, and they do share a cognitive profile of difficulties, that, even though they can be expressed in very diverse ways, there's still something that makes all of these individuals autistic, in a sense.

This, again, I read as insistence on qualitative distinction—an unchanging core, a *something* that is remarkably indifferent to heterogeneity and complexity. I said: "So it's not just an umbrella of convenience, then." And he replied:

> I think if you talk to pretty much, you know, generally the scientific community on autism, you will kind of—they will pretty much all tell you that you know whether somebody is autistic or not. So there's a certain kind of feel to the interaction, and you just . . . it takes a bit of time, once you've met a certain number of people with autism, you just kind of develop a radar for it.

Again, the word used here is *feel*—understood as a perfectly good "way that you know" about autism, within the "scientific community." Later on, I asked the same person about a hypothetical "brain scan" for autism and whether he would be inclined to rely on the brain scan for diagnosis or on the *feeling* of an interaction. He paused for a bit. "That's a good question." He said eventually:

Probably the interaction—because autism in the end is defined by a collection of behavioral manifestations. So, if I put you in the scanner, and
your brain looks like an autistic brain, and then you behave in a completely
nonautistic way, I wouldn't call you "autistic." Whereas if your brain wasn't
autistic, and you behaved in a very autistic way, I'd probably think that you
were autistic. So if the two are in conflict, then you always kind of go with
the behavioral manifestations, and those are the ones you pick up in the
interaction with somebody. So I'd probably go with behavior over brains, or
biomarker.

I am struck by the way in which the relationship between a collection of
behaviors, and such a definite entity as an autistic brain, gets negotiated
by a reliance on, and deferral to, the obviously affective labor of *feeling* the
quality of an autistic interaction. The sense of feeling here grants the scientist a simultaneous commitment to the "collections of behaviors" *and* to the
"something" that "makes someone part of the autistic spectrum." Something
similar emerged in a conversation I had with a much more senior researcher,
well known for her input to debates around the "heterogeneity" of autism.
When I went to interview this professor, I had expected her to expound
more on this theme, but even though she did talk through this topic, I came
to realize, as the conversation developed, that she was more keen to stress
that autism was something that nonetheless remained, in her words, "true to
itself." One of the ways she had come to recognize this trueness, she pointed
out, was through recognizing autism in other cultures. She said:

When I went to Japan, I understood the autistic culture much more than
the neurotypical culture—and could recognize much better where the
questions that the person with autism was asking me came from. And the
colleague I met, who I think had Asperger's syndrome—I understood more
about maybe what his view of the world was, and what his expectation of
my behavior would be, and what he wanted from me, than what the colleagues who'd invited me . . . [with] who[m] I wasn't sure if I was maybe
doing the wrong thing or, you know, stepping on toes. . . . And then there's
the element of seeing how neurotypical kids start off very idiosyncratic—
very funny, and very much themselves. And then they get sort of . . . it's
not really peer pressure, they just become part of the herd. And children

become less interesting, in the sense that they become more predictable—
you know, you can say what most children will be interested in by a certain
age [*inaudible*] so there's a kind of a blinkers put on through socialization.

Here is a sense of autism as something that can be known precisely because
of its qualitatively distinctive *in itself* qualities. The specific quality of autism
is not marked by a *feeling*, so much as a kind of tacit, intersubjective *recognition*—in areas (such as other cultures) where, if autism was just a matter of
diagnostic convenience, you might expect it to look different. And this is not
at all a question of *whether* or *how* autism is diagnosed in other cultures but
rather of whether the individual who had learned to see it, to feel it, to recognize it, to interact with it somewhere else . . . whether or not she can see,
feel, recognize and interact with it there too. Another senior professor to me:

> We know that autism is heterogeneous in terms of aetiology. So all the
> individuals who have what we call "autism" won't have it for the same
> reason; the cause won't be the same always—[although] in some it might be
> a pathway that overlaps or is common to some presentations. And we know
> that it's heterogeneous, in regard to [the fact that] different people who
> have autism can look very different from each other, and individuals change
> over the life course a lot. So there's all this heterogeneity. But we also think
> there's this thing we recognize when we see it, and it's this thing called
> autism.

This is precisely what I have been trying to draw attention to—that we know
all about the heterogeneity, the complexity, and the different causal pathways. Yet despite all of that, there is something we know when we see it—
and this something is autism. This is an almost embodied, affective idea of
autism as something that, for neuroscientists, manages to transcend both
the laboratory space of biology *and* the clinical space of diagnosis; in these
accounts, autism is simply known and affirmed, when it is seen, or felt, or
recognized.

This commitment to feeling and recognition is, I suggest, one of the
most important registers that neuroscientists draw on in order to trace
autism through its tricky and crossed appearances. It opens up the story of
how autism gets traced through a research program as *both* an unchanging

reality *and* a convenient way of linking some genetic and environmental nodes. Thinking with the capacity and willingness of neuroscientific researchers to talk about autism as an enigmatic object of fascination, and also as a felt or a recognized "thing," opens up a new perspective on how neuroscientists conjure the objects of their attention. It represents a first step in figuring out how those objects are sustained through such heterogeneous and unpredictable fields.

WHAT'S AT STAKE

It would be easy to see the emergence of autism through the prism of biomedicalization or neurobiologization—and to take the neuroscience of autism as just another step in the long march of neurobiological imperialism. Here, after all, is a relatively new diagnosis, one heavily bound up with emerging technologies and with the marketization of health; the full conceptual apparatus of neuroscience has since been brought to bear on it, including on how we define it, and how we understand it; it is also a diagnosis in which we could clearly see a shift in the medical gaze away from categories of discipline and cure—and toward the maintenance of health, the management of chronicity, and the endless sifting for as-yet-unknowable genetic and neonatal risk factors (Rose 2007).

I have tried to look in depth at just one part of the complexity that lies beneath this story, a complexity marked by the inability of neuroscientists to even agree that autism is much of a meaningful biomedical category in the first place and also by their willingness to find themselves sometimes a little dumbstruck by the complexity of it. None of this is intended to particularly run against the convincing framework put forward by Adele Clarke and her colleagues (2003: 184), which is very much alive to the "contradictions and unanticipated outcomes" with which this vast, ambiguous shifting complex is inevitably associated, and there may well be room in their account for the way that my interviewees talked about autism. What I am trying to show, nonetheless, at the coalface of neuropsychiatric biomedicalization, is the complexity and ambivalence of the process through which a diagnosis actually becomes entangled in the machinations of bioscience and biomedicine.

There is an important task for the social study of biomedicine in understanding these scientists' ability to trace their practice across an often tricky two-way commitment to heterogeneity and distinction. As far as the new brain sciences are concerned at least, and particularly as they run into developmental diagnoses, the path to biomedicalization might not always run very smooth; it even may sometimes run *so* unsmooth as to no longer leave us in great confidence about the destination. We still need to understand how categories of practice, epistemology and affect, can explain what animates such a tricky and sometimes contradictory development as biomedicalization or neurobiologization. Certainly it seems to me that these delicate and unexpected registers, like feeling and fascination, that help neuroscientists to carefully trace organic phenomena through, for example, point changes in the brain, pattern changes, parametric vectors, and pattern-classifications, should probably play a more prominent role in discussions of neuropsychiatric and neuropsychological research more generally.

All of this matters too. Although I have presented this, for clarity, as a discussion taking place within clinics and laboratories, in fact many people and groups have a stake in the "biology" of autism and in the specificity of the relationship between autism and the brain in particular. On one level, there is the politics of parent activism. Ever since the psychogenic account of autism located the disorder in a kind of maternal coldness, autism research has been beset with a familial and gendered neuropolitics and an intricate politics of expertise, which have together formed a series of decades-long contests in which the mantle of "science" has been (successfully) claimed by formal amateurs and in which laboratory-based biological knowledge has proceeded in alliance with parents and their advocatory organizations (Feinstein 2010). Unquestionably, the kinds of oscillating accounts that I relate in this chapter, and the uneasy movements between "biological truth" and something more disparate that I have described, are structured in part by this politics. But there is another politics at stake here, and this is the politics of neurodiversity, which draws on the rhetoric and strategies of older identarian and liberatory movements to stake a claim for autism as a difference to be respected and not a disorder to be cured. This claim—newer and less mainstream than the familial politics just described—is even more heavily invested in a more or less solid identification of autism with some unimpeachable neurological substrate (Silverman 2011: 163; Ortega 2009: 434). I have presented here a kind of academic and quasi-philosophical back-and-forth about biology, diagnosis, convenience, enigma, and so on. But it is vital to recognize, in conclusion, that there is nothing trivial about how autism gets traced though both an organic and a diagnostic-convenience view, still less how successful and convincing such a delicately traced neuroscience can continue to be. Within the difficult politics of disability and difference, this question really does matter.

2

THE TROUBLE WITH BRAIN IMAGING

I N SOME WAYS, THE KIND OF UNCERTAINTY ABOUT AUTISM DISCUSSED
in chapter 1 was predictable—the autism "enigma," after all, is one of the
defining tropes of the field. But what was a good deal less predictable was
the amount of ambiguity I encountered, from neuroscientists, about *neuroscience*. Not invariably, and rarely without qualification but still frequently
enough to be unignorable, my conversations with autism neuroscientists
were braided with intense feelings of doubt about neuroscience itself. This
was a surprise. I had expected these scientists to be, as a cohort, more or less
confident about what it was they were actually doing. Instead, I discovered
a set of individuals who had markedly different views—and those views
sometimes strikingly downbeat—about neuroscience as a set of practices,
about how useful they found it, how much faith they had in it, how reliable
they considered it, and so on.

Not all subscribed to this register: for some, the methods and rubrics of
the new brain sciences were, as I had anticipated, a mark of disciplinary and
epistemological strength, the sign of a maturing science of mental disorder,
and a cause for clinical and therapeutic hope. Those more optimistic contributions tended to focus on the excitement of using new tools to identify a
biomarker for autism as well as the hope of uncovering autism as an essentially organic phenomenon, amenable to intervention. But there was also
(sometimes even at the same time) a strong current of disappointment running though these interviews. There was sometimes even a sense of something close to *anxiety* about the methods, assumptions, and technologies
inherent to the new brain sciences. For example, quite a few people talked
about their own practice within a register of distinct unease about the arti-
factual nature of brain-imaging methods; some expressed a discontent with
these methods' lack of objectivity; others talked about their relatively small

contribution to psychology as such; for others, it was about the presence of "blobs," the ways that data can get manipulated, and so on. Just as the neurobiological account of autism was traced through mutually exclusive-looking registers of biological truth and diagnostic convenience, so did my interviewees exhibit a strange and dynamic relationship to neuroscience as a practice. This was a relationship that was as well versed in the language of disillusion and disappointment as it was in the straitlaced semantics of hope.

EXPECTATIONS

The dynamics of expectation now constitute a fairly well-established field in the social study of the life sciences. Most prominently, the sub-field known as the "sociology of expectations" (Brown and Michael 2003) has directed attention to the way that emerging scientific ventures are often maintained by, or oriented around, discursive structures of hope, optimism, and positive expectation (van Lente and Rip 1998; Borup et al. 2006). However, more recent work has nuanced—and even troubled—this rubric. The sociologist Martyn Pickersgill (2011b), for example, has drawn on his interviews with frontline scientists and practitioners to argue that, within the practice of neurobiological research, there might be rather more uncertainty, and rather less positive expectation, than has previously been identified (cf. Moreira and Palladino 2005; Nerlich and Halliday 2007; Tutton 2011). More recently still, John Gardner and his colleagues (Gardner, Samuel, and Williams 2015: 1004), in a study of deep brain stimulation, have proposed a broader role for *low* expectation in biomedical innovation. They argue that, far from being anomalous, such registers play "a vital part" in helping clinicians to negotiate uncertain therapeutic landscapes. There is much to be said here, but I am going to stay at a bit of an angle to these debates in what follows. I am with Gardner and his colleagues in trying to think more generally about what it is that something like disappointment, or at least the potential for disappointment, can *do* in clinical and research settings. But in pursuing this question, I cleave more closely to my own interest in what gets done in the midst of complexity and ambiguity in neuroscientific practice—to the constitutive and generative work that gets *produced through* uncertain, and sometimes deflated, ways of relating science.

This chapter is about how neuroscientists tack back and forth between, on the one hand, an excited and optimistic genre of talk, sometimes even fading into a techno-utopianism; and, on the other, an emergent sense of uncertainty and ambivalence about what it is that brain-imaging technology actually *doing* here. I suggest that the copresence of these contradictory accounts does not only reveal an unexpected complexity in how neuroscientists think about neuroscience: rather, and pulling my interest in tracing now through an engagement with feminist theorist Karen Barad's "agential realism" (2007), I ask if we should not in fact be quite *unsurprised* when a scientific practice gets talked out through complex, equivocal genres of disappointment and hope. I ask if we should not be especially unsurprised when that science is characterized by its commitments to, on the one hand, the complex entanglement of scientific work and scientific objects (although I don't like the word "objects") and, on the other, the biological singularity (I don't like singularity either) of that object *all the same*.

Two commitments that I flagged in the introduction are also going to become more obvious along the way. First, although I am working very hard indeed to avoid my account being read as any sort of contribution to a "turn to ontology" within studies of science and technology (see, e.g., Woolgar and Lezaun 2013), I cannot pretend that these conversations are free of ontological commitment. Thinking with Karen Barad, however, allows me to unspool some of those commitments while sticking closely to the accounts of the scientists that I am in conversation with—and especially to set out, in more detail, what is actually going on in the strange simultaneity of confidence and uncertainty that peppers their talk. Second, in this chapter I offer some comment on the practice that, in the social study of neuroscience, names itself "critique" (Choudhury, Nagel, and Slaby 2009; Campbell 2010; Kirmayer 2011). Toward the end of the chapter, I ask: What is the future of critique, as it orientates itself to a scientific practice that is *already* so downbeat, so troubled, and so anxious? What can the critical task of an interpretive sociology of science *be*, when its interview subjects are so insistent about—and so skilled at—self-criticism? What happens to a critical sociology of science when the most stringently critical voice is that of the natural scientist herself?

THE DREAM IS TO INTERVENE

First the positive stuff: throughout the research for this book I found that when my interviewees talked about their orientation to neuroscience particularly, their talk was often shot through with rich discourses of hope, possibility, and expectation. Of course the study of mental health has often been quite formally structured by a sense of clinical or therapeutic hope for the future (Moreira and Palladino 2005). For neurodevelopmental problems this hope has recently become embedded in, and articulated through, the search for brain-based biomarkers particularly and the emergence of novel neuroscientific technologies that might mark these out (Raff 2009). In the first pages of the first issue of *Nature* published this decade, for example, the editors self-consciously framed the 2010s as "a decade for psychiatric disorders," a sense of optimism and expectation that was quite precisely rooted in the idea that "new techniques—genome-wide association studies, imaging and the optical manipulation of neural circuits—are ushering in an era in which the neural circuitry underlying cognitive dysfunctions . . . will be delineated" (*Nature* 2010: 9). The authors went on, citing Thomas Insel, then head of the US National Institutes of Mental Health (NIMH): "Whether for schizophrenia, depression, autism or any other psychiatric disorder, it is clear . . . that understanding of these conditions is entering a scientific phase more penetratingly insightful than has hitherto been possible" (ibid.).

My interest is not in how accurate this claim is (although the ensuing decade certainly proved more awkward for neuropsychiatry than these authors anticipated (see Kapur, Phillips, and Insel 2012). Instead, I am only focused on the degree to which these kinds of hopes, common enough in such a public-facing literature, were also mirrored by the frontline researchers that I interviewed. One researcher said:

> What intrigued me in the early days about MEG [magnetoencephalography, a form of brain imaging] is that, first of all, it is a beautiful combination of quantum physics, which is the underlying principle of the scanner, and the application to not only biological, but human, and even psychiatric problems, or neurological problems. . . . it was sort of immediately a very sort of

appealing way of having the dynamics of the human brain measured with a tool which is capable of capturing these dynamics.

This view, that technologies like MEG would open up the human brain and give new insight into psychiatric and neurological problems, was not uncommon: "All the neurology-type people are looking for the biomarker, you know," one professor of psychology told me, ". . . and I think they have implicitly in their heads this notion that we will find something which will then [*whooshing noise*], it'll part like the Red Sea." Or as a former employee of a research funder put it:

> There was a very strong sense about five, six years ago . . . that the technologies to create the breakthroughs in conditions like autism were coming through—the neuroimaging technologies, the genetic analysis technologies, you know, and the sort of bringing to bear, if you like, of those technologies, you know, the sort of access to brain material and the kind of imaging that you could do with brain material, and indeed the chemical procedures that you could do with brain material. And that's proved to be true. Things have moved forward enormously in the last five years—to the point where forms of intervention that are based on biology are now feasible.

For researchers, what this technological hope ultimately expresses is the prospect that the field will significantly advance in some way. And in psychological and psychiatric research on autism, this goal often manifests as an expectation of reducing the field's reliance on behavioral measures for diagnosis. This came up quite forcefully during a conversation that I had with one researcher—herself involved in innovative work to find a quantitative-organic marker for autism. Her background (perhaps tellingly) was more embedded in MRI analysis than it was in psychology. She told me about her experience of joining her current research project on autism and being trained to use a behavioral autism scale. She said:

> I was amazed at how many details these people [the trainers] pick up on, like, you know, you speak about instrumental movements and so on, goal-directed actions, and I just couldn't see it. And I could only do it with a lot more training—I'm talking months here.

Her amazement at the skill required by behavioral analysis is not only a compliment to clinical skill, I think; it was also expressive of a more fundamental concern—that is, that such skill is required at all. There had to be a better, more predictable way to go about this. And the best hope for advance lay with the new brain sciences. She said:

> If you look at the behavioral studies, there are not too many differences on the behavioral level, when you look at adults. But there are also a few brain studies now coming out that show, actually, in terms of their anatomy, people with Asperger's [syndrome] are different from people with high-functioning autism. . . . if I was a behavioral researcher, I would feel that that [behavioral research] has maybe come to an end, because if we are now speaking about, actually, Asperger's or HFA [high-functioning autism] is the same behaviorally, what are we going to research on—what comes next?"

Her basic hope, in other words, is that developments in brain-imaging technology will reveal a difference in brain anatomy (between autistic and "typically developing" people) wider than the difference in behavior—delineating and marking autism at a finer level than is currently possible for even the most skilled clinician. "Biomarkers could not only reveal causes of the condition," as the ethicist Pat Walsh and her colleagues (2011: 603) note in a careful review, "but could also be clinically useful in complementing or improving the behavioural diagnosis of autism and in enabling earlier detection of the condition."

A second and related technological hope is that with the disorder locked down to an identified anatomical pathway, a clear entry point for intervention is opened up. As the scholar of neuroscience and society Ilina Singh has pointed out, in a paper with the sociologist Nikolas Rose (2009: 202), this sense of opportunity is characteristic of contemporary psychiatric-biomarker research in general but is particularly acute within the realm of the neurodevelopmental disorders. Hopes of diagnosis and treatment in turn are premised on another expectation—that if clinicians could intervene on the neurological substrate before behavioral symptoms appeared, this would likely prove more effective and more efficient in the long term. One senior professor said to me:

> People with autism have got biological differences in brain development, you know, so that that's . . . and that's related to some of the things they do. The

thing that we're work . . . and we've just identified what those differences are [using brain imaging]. And we're in the middle of saying, "Can you use those differences to diagnose people with autism rapidly, and/or in a cost-effective way?"

This was a common theme: "The dream is to intervene prior to the onset of symptom," said one leader of a major brain-imaging project. "You know, to try and divert the developmental pathway before the full core symptoms of autism become kind of embedded in the system." Discussions of the desire to wring early diagnosis and treatment from neuroscience were never unsubtle in my interviews, nor were they always present. But they were there all the same—and frequently so. Perhaps summing up this view, one senior professor put it to me like this:

> I don't think there are many people in autism who would say that they don't want to understand other people, even if they choose not to engage with other people at the level, so I would expect that [one day] we would be able to intervene psychologically, neurologically.

The idea of treatment is controversial in autism—and particularly so the idea of intervening neurologically at a very early stage (Barnbaum 2008; Barnes and McCabe 2012). I didn't encounter anyone who was unsympathetic to the view that there are good reasons to be wary of such interventions, but I nonetheless encountered, frequently enough, the hope that people with autism would be able to be diagnosed earlier and treated in the future, specifically by acting on the brain. As the psychologist Laura Schreibman (2005: 133) has pointed out—directly after acknowledging the "controversies *within* controversies" that structure this debate—"we still have no cure for autism. Yet there is reason to be hopeful." I encountered many neuroscientists who carried this hope.

NOW WE HAVE ALL THESE WONDERFUL TOOLS

Quite a few scholars of science and technology have turned their attention to thinking about the role of hope and expectation in gathering together large-scale, diverse technoscientific projects—such as the search for a brain-based biomarker of autism—and have begun to identify some of the ways that these projects actually get justified and assembled in the present, through the expression of some promise or prospect for the future (van Lente and Rip 1998; Brown and Michael 2003). Scholars within this tradition have focused on the role of the specifically *promissory* "expectations" that are often attached to scientific and technical projects and around which resources and actors can begin to assemble themselves: "technological futures are forceful," Harro van Lente (2000: 59) has pointed out. "Once defined as promise, action is required."

By "expectations" they mean "wishful enactments of a desired future . . . hyperbolic expectations of future promises and potential" (Borup et al. 2006: 286). The emphasis is mostly (although not entirely) on a scientific desire to imagine something basically *good* for the future, through the assembly and propulsion of some scientific and technological practice. On the basis of this promise, it becomes reasonable, even imperative, to enact that project or practice in the present. As the sociologists Nik Brown and Mike Michael (2003: 4) have argued, through the articulation and enactment of varieties of expectation, the epistemic and practical distance between the past and the future is discursively (if not materially) elided: "The future is mobilized in real time," they point out. Across this narrowed gap, elements of a research program can be rapidly drawn together.

Unquestionably, this "sociology of expectations," although usually focused on more public discussions (Kitzinger 2008), at least partly explains what's going on in my conversations—insofar as these expressions of hope can also be read as the outline of an assembly practice. In this sense, loose promises of neurological diagnosis and therapy in the future become the ground on which large-scale projects are enacted in the present. Brown and Michael (2003: 12–13) have noted an inverse correlation between closeness to the actual scientific practice and the level of expressed hope. While I found these expressions at all levels among the scientists in my interviews, from PhD students to senior professors, some of the more compelling and thought-through articulations came from the (slightly more distanced) leaders of large-scale projects, who were clearly not articulating their sense of hope for the first time and for whom a convincing image of expectation likely played a more directly instrumental role.

For example, when I asked the principle investigator (PI) of another large project about why exactly someone like him—a prominent neuropsychiatrist with diverse interests—would actually pursue something as awkward-seeming as a neuroscience of autism, he imagined precisely the kind of promising-future scenario that the biomarker discourse is organized around. He said:

> Say you go into [the] accident and emergency department with a cardiac arrest. Now, option A: you describe to me your symptoms. Crushing chest pain, burning sensation going up into your neck, pain coming down your arm, right? Feeling sweaty. Not feeling chipper. And I say to you, "Oh, really? Sounds like you might have something going on in your chest." But you would expect me to do an ECG [electrocardiogram] to measure the function of your heart, right? Or if you went in there thinking "I've got diabetes," you'd expect the doctor to measure your blood-sugar, right? If you went in there with epilepsy, you'd be expecting him to measure your brainwaves. Well, why should you not be doing the same thing if you go in with a biologically based neurodevelopmental disorder? I want to be measuring whether you've got an abnormality in the organ in question. [. . .] If you think there's an abnormality in an organ that's causing a behavioral difference or behavioral abnormality, you've got to measure what's going on in the organ.

We can see here, in fairly bald terms, the basic hope of diagnosis and treatment that is invested in brain-based biomarker research and around which that program has become organized—that is, the hope that neuroscience will one day make autism as instantly and definitively diagnosable as a heart attack. But what is also interesting about this imagined scenario is that it plots both backward and forward in time, to argue (and from memory, rather forcefully) that the basic promise of a neuroscience of autism is to provide access—conceptual and methodological—to the organ that researchers had really *always* been investigating but to which their methods, up to now, had simply been inadequate. This, I think, is at least one part of what a discourse of expectation can do for the autism researcher particularly, which is to make sense of an awkward past and present in the light of some visionary future (I have more to say about the relationship between science, memory, and especially psychology in chapter 4).

Essentially the same view was expressed even more bluntly by another senior professor, who sat at the apex of a fairly large program of research. He said:

> I think neuroscience always believes that psychology was always a sub-part of neuroscience, but in the 1970s and 1980s within psychology there was a very, very strong push to, you know, not be misled by data from neuroscience. And I think it's, you know, partly a theoretical thing, partly a methods thing as well—because we didn't really have the methods, other than looking at patients with a very messy brain hemorrhage which wasn't very, you know . . . or doing animal studies. Now we have all these wonderful tools for functional imaging of the brain which we didn't have in those days.

Another told me how, today "a lot of psychologists have redirected the focus of their work onto looking at not just the cognitive basis of some kind of process like memory, or attention, or in my case, social cognition—but also the brain basis."

The "sociology of expectations" tells us that this is not an empty optimism and that its expression actually helps to assemble the various elements of the biomarker research in question. I found this at all levels within my interviews but expressed with particular clarity and force by the project-leader scientists for whom the work of project assemblage is clearly very direct. It is perhaps no surprise to find the presence of these expectations within autism neuroscience especially, a notably awkward area of biomarker researcher, one known for its troubled past—an area of research where expectation and future fulfillment may have especially prominent roles to play. But what was more interesting about my interviews with these scientists is that hope, optimism, and expectation were only one part of the story. At the heart of these conversations about the relationship between neuroscience and the autism spectrum, there was also a strong current of unease, disappointment, and even some anxiety about the developing program of research. This moves us quite sharply away from the "expectations" literature. It begins to tell us something different about the neuroscience of a category like "autism" and about the ambiguous terrain in which it is suspended.

A VERY INDIRECT MEASURE

Despite prevailing popular and media sentiments about "the rise of neuro-everything" (Vidal, cited in Rapp 2011: 7), the urging of restraint and explanatory parsimony as well as the routinization of sharp internal critique is actually a recognizable feature of the public discourse of neuroscience (Logothetis 2008; Vul et al. 2009).[1] In 2016 it is no longer such a surprise to see a class in a psychology department titled "Everything Is Fucked."[2] But I had still not anticipated the sheer volume of negative sentiment about neuroscience, especially imaging neuroscience, and what it could or could not tell you about autism, that I heard throughout the course of this project. Indeed, and despite my own self-consciously bland and uncritical presentation, interviewees from cognitive neuroscience consistently, and often with some vehemence, drew my attention to the problem of false positives, the distance between what their methods measured and what they purported to measure, the degree to which neuroimaging simply replicates what is already known through other means, and even the basic inadequacy of brain imaging to phenomena like autism in the first place.

I am not claiming that any of these issues are shocking or unknown. But I do want to say there is more at stake in these conversations than an appropriately scientific caution. The consistency and depth of these cautionary anecdotes that I encountered among autism neuroscientists who were often, themselves, positively expectant too must complicate discussions about the role that a bullish sense of promise plays in neuroscience more broadly. Consider, for example, the following account, which is about fMRI (functional magnetic resonance imaging) neuroimaging of autism in particular. It comes from a young autism researcher, whose intellectual and methodological hinterland was actually more in a hard-nosed cognitive neuroscience than it was autism research or psychology as such. She said:

> You've got to be careful with neuroimaging and the questions you ask, because the problem with neuroimaging [is that] you'll always get a result—you'll always get some blobs, you know? [. . .] I always say, I used to laugh to people and say, "Oh my God, this is an art, not a science" [*laughs*] because . . . you've just got to be so careful. And I think there's a real truth

to neuroimaging. I believe in it. But it's one of those things that require replication—and the truth will out, and if you've done forty studies on social cognition and thirty-eight of them are showing the superior temporal sulcus, then I think you can hold your hand up and say, "Well, this area is involved in social cognition," which is really important, but there's a hell of a lot of other blobs, and that's not a very nuanced finding either [*laughs*], it's a bit crude, so I think to get . . . I think it's got a long way to go, and people have got to be really careful.

I particularly want to note here the tension between this neuroscientist's commitment to the basic truthfulness of the image and, nonetheless, her acknowledgment of how heavily mediated the process of production is, and how much artifice is potentially involved in the interpretation. Although she finds some resolution in urging care, and also in deferring to replication, the nervous laughter and also the anxious doubling back of the account even when it seems some basic resolution had been achieved ("it's a bit crude") suggest to me the presence of a more ongoing concern.

In her ethnography of MRI imaging, the sociologist of science and technology Kelly Joyce (2005: 438) draws attention to some of the "rhetorical practices [that] produce a construction of MRI in which the image and the physical body are seen as interchangeable." In particular, and especially in clinical settings, Joyce (ibid.: 458) has argued that popular discourses about transparency, revelation, and truth often obscure the manifest variability and unreliability of MRI brain imaging—even in discussions between clinicians and technicians who are quite aware of the limitations of the method. "Wow," said one physician to Joyce, when talking about his relationship to MRI, "it's as if you sliced a person in half and looked at them" (ibid.: 437). Granted that it is a much more epistemologically compromised practice, but quite a few of my interviewees, usually using functional MRI, expressed an entirely different view. Indeed, rhetorical practices among my interviewees *repeatedly* constructed brain imaging as something that was artificial, unreliable, or even manipulative. "It bears no relationship to reality," said the highly regarded leader of one laboratory to me.

In particular, technical problems with the generation, processing, and handling of brain-imaging data were repeatedly forefronted. The following extract comes from a psychiatric neuroimager who had worked on quite a few autism projects but who also, and perhaps even more so than the person quoted earlier, was intellectually embedded in the hard science of MRI and fMRI analysis. Lamenting the generally weak understanding of the physics of these technologies among psychiatrists and psychiatric researchers, he drew particular attention to the phenomenon of resting-state data (Raichle et al. 2001).[3] He said:

> The resting-state data came about because people started thinking about so-called deactivations, and noticing that these deactivations were appearing in virtually every data-set. And people ignored them. People literally airbrushed them out of their results. They just didn't want to know.

I was struck, at the time, by how scathing this interviewee was about people's use of the method ("They just didn't want to know") and also how irredeemably problematic he found the method in general ("virtually every data-set"). This is not a story about the need for proper scientific caution; nor is it a story about the basic scientific pragmatism and scepticism that override a nonetheless tempting narrative about the objectivity of brain imaging. He said later:

> The thing about science in general is that what counts is money and real estate. . . . So, people in offices give you power, give you influence. As does grant money. And the two things tend to go together. Again, if you're just sitting in an office writing things down on a piece of paper, that might be great research, but doesn't necessarily bring in much income. What brings in income is doing big studies that employ lots of people, then those people become dependent on your goodwill, and so then you have influence on them. And so obviously that's the way it works. The huge increase in scanning, of course, people are thinking that would be a way to get power and influence by, you know, bringing in research money and so on and so on. Well, that's what happened.

Although this situation of technology, and the desire for technology, within the political economy of contemporary science and the academic politics of the university is probably not so rare, it is striking to have this view narrated through the large-scale advent of brain scanning, particularly. Moreover, the generally deflationary approach to brain imaging that he expressed was not at all unusual. "Brain imaging is based on a lot of assumptions," said a postdoc to me, with a sigh. "You know you must be measuring something in the brain . . . but it's correlates of that thing." Another researcher expressed the same view: "fMRI is a very strong [technology] . . . but it is a very indirect measure."

I am not pitching autism neuroscience as a contradiction or a counter-example to the "sociology of expectations" literature, which acknowledges both that "expectations" are not always positive and also that even positive futures will generally coexist with some sense of failure or simply frustration. As Borup and his colleagues (2006: 290) have put it: "Disappointment seems to be built into the way expectations operate in science and technology."[4] Yet still it seems to me that the roles of deflation, anxiety, and uncertainty are not emphasized enough within this corpus. That still, by and large, when we talk about the sociological import of "futures" and "expectations" in scientific project-making, we are talking about actors orienting themselves to something they imagine to be basically optimal. The future in question still tends to be one in which: "Gene therapy and nanotechnology will cure disease, cars will drive themselves, pigs hearts will be used for organ transplants, computers will become an even more ubiquitous part of life, the Internet and the Cybercafe will become the venue of choice for our relationships, and so on" (Brown, Rappert, and Webster 2000: 4).

What we see in the closeup space of neurobiological autism research, however, is a significantly more complex and dispersed terrain of expectation—one that works through some notably deflationary, uneasy, and even quite disappointed views of its own basic project. It is not enough to describe this data as an undercurrent of knowledgeable scepticism within everyday research. This sense of disappointed uncertainty was too present and too much a feature of my interviews with autism neuroscientists. My question now is whether, if we want to understand the intellectual landscape of the contemporary neurosciences, we might need to stop thinking with promissory futures and start focusing on the ways that scientists mobilize a constitutively *ambivalent* relationship *both* to languages of uncertainty and to discourses of hope (Silverman 2011: 159–60).

THE TROUBLE WITH BRAIN IMAGING

About halfway through this project, I met a young researcher who was contributing to several major brain-imaging studies of autism (although, notably, more with EEG than fMRI measures). Having come to brain imaging from biology, she was keen to express her early disappointments in this field. She said:

> When you know how the brain works as a biologist, so you know what makes brain activity, which is connection between neurons, and it matters with which part of the brain you're connected, and how fast you get there, and how much information you converge . . . um, the only thing you get from brain imaging is "this part of the brain is activated at a particular time." It tells you very little about the neural mechanism and how things get connected to each other.

Here, the interview shifts from a basic concern with the distance between brain activity and some measurable vascular response to a more specific and profound worry about the relationship between the kind of data generated by brain-imaging measures and, in general, "how the brain works." I want to draw attention, in particular, to some important normative divisions that are being enacted—between brain imaging and biology, on the one hand, and even between brain imaging and brain *science*, on the other. A senior molecular biologist who worked on autism said something very similar, but she embedded her qualms, not so much in the way that connection was being elided but in the degree of fineness achievable from brain imaging. She said:

> In autism, along with a lot of other conditions, like schizophrenia and even the neurodegenerative conditions, you really need to understand what's going on with gene expression in the brain. . . . The trouble with brain imaging is that it only gets you down to a certain level of fineness in its detection. So you can't tell what's going on at the cellular level, and at the molecular level—which is what you really need to understand if you're going to see what the genes are doing, and what it might be possible to do to improve symptoms that some people with autism have.

On one level, we could read here a wet-lab biologist's anxiety about the oft-remarked "seductive" nature of the brain image and the degree to which the brain is thought to have an intuitive appeal through its visual relationship to some notion of organic truth. As the neuroethicist Martha Farah (2005) has pointed out, people tend to "view brain scans as more accurate and objective than in fact they are" (cf. Weisberg et al. 2008). In his *Picturing Personhood*, anthropologist and STS scholar Joseph Dumit (2004: 6–7; emphasis in the original) has argued that "there appears to be something *intuitively right* about a brain-imaging machine being able to show us the difference between schizophrenic brains, depressed brains, and normal ones." The interviewee just quoted said: "This tremendous emphasis on imaging . . . has led people to think that everything's virtual these days when, actually, it only gets you a certain way, that virtual reality."

It is interesting, though, that the focus of her unease is not an over-interpreting public discourse; her worry is about the limitation of a brain-imaging study of something like autism in the first place. This sense of limit, which was one of the most consistent sources of unease and disappointment expressed about brain imaging within my interviews, came out in a few different ways. For some, it was about thinking small. As one young psychiatrist said:

> In some forms of research I suppose you might come up with a finding which sort of clearly changes the game. And in brain imaging in autism, it's rarely that sort of finding. So the findings usually sort of move things on in very small steps.

For others, it was about recognizing ineffable complexity: "I don't think it's ever going to be as simple as, 'there is this point in the brain that is dysfunctional and this is causing autism,'" said a PhD student. "I don't think that's ever going to happen. I don't think that's true." For one of the senior investigators already quoted, the problem was lack of specificity: "Quite a good pub game," he said, "is name a region of the brain that hasn't been associated with autism, by somebody or some paper. It's virtually impossible." Of course, these are not suggestions that the neuroscience of autism is intrinsically bad or misguided. But there is a subtle but consistent sense in which the neuroscience of autism is described as limited small-scale, dispersed, and (so far, at least) not very specific. Whatever hopes had been attached to their research, these scientists expressed some quite consistently *low* expectations for the neuroscience of autism. None of them thought that this meant neuroimaging research shouldn't be done (and several went on to talk about the move to "connectivity" or some other new paradigm. See Wickelgren 2005; Anderson et al. 2011). I was nonetheless struck by the way in which the neuroscience of autism was consistently self-constructed through an idiom of uncertainty, one that emphasized the biases, the difficulties, and the partial truths.

These are not just aberrations or counterexamples of a broader structure of hope, nor are they the predictable post hoc sentiments of people whose research hasn't worked out. It is also important to emphasize that low expectations do not correlate either with disappointed careers in my sample, nor are they particularly found among the junior and the put-upon: all those just quoted are "successful" scientists by any reasonable measure. My point is not that I have found scientists who are unhappy or drifting, or who find themselves inadequate; nor is it the case that I have interviewed comfortable field leaders, whose long-established sinecures let them give free range to their doubts. My point is that analyses of "promissory hope" in neuroscience may need to be tempered by a greater attention to more modest visions— that among this set of interviews with researchers who work on autism, who principally conduct their research through neuroscience, the work of actually putting this neuroscientific account *together* gets traced as much through a sense of unease about how a neurobiological autism might come about, and what it would look like if it did, as it is structured by a sense of hope for this practice in general or for the therapeutic and diagnostic hopes it may yet realize.

Perhaps most damningly for a practice that lives or dies on its sense of efficacy, there were also suggestions that brain imaging will only ever go over ground already well-trodden by other experimental psychologies. A lecturer in cognitive psychology said:

> I don't necessarily see brain-based work as an explanation as such. So I think people . . . it's . . . to me, it kind of adds a layer of description. So, for example, this is a very simple example, but say we're taking about face-processing and I say that children behaviorally have difficulties processing

faces. And you can do tests to show this. And then, at the neural level, they show less activity in the fusiform face area for faces. To me, that kind of is just another level of description. It doesn't explain anything.

Reading this again, I think this lecturer—someone who has been involved in quite a few brain-imaging studies—is quite deliberately trying to enact a firm division between cognition and the brain; she is positioning them as different (even competing) areas in which to seek the most richly explanatory substrate of a given mental state. Furthermore, she is suggesting that neuroscience doesn't necessarily "add value" to what we already know.

But aside from disciplinary or institutional positioning—and that at least explains in part what is going on here—this view expresses another element of this general neuroscientific self-critique, which is that autism neuroscientists have been scanning brains for about two decades now, yet it's not clear that the field has dramatically moved forward. "I think looking at the brain is useful in some respects," said a young postdoc echoing this view, "but, um, I mean I am always saying that I think a lot of sort of neuroscientific work, especially in terms of fMRI or stuff like that, is a process of redescribing what we know already." Or as another lecturer put it: "I did see a talk here recently on—it was called 'the neuroimaging of ADHD' and that was what it was. And of course functional neuroimaging by itself is meaningless. Because it is just lighting up pictures."

When these researchers say things like "it's meaningless" or "it doesn't explain anything," they are articulating (I think) a basic anxiety that there has been a collective overinvestment in the brain—and that, in fact, attaching categories like autism to localized neurological signatures might not add a great deal to the field. This is where I think we reach something close to the opposite of a promissory vision. Indeed, there was even a suggestion within my sample that the opportunity cost of neurological redescriptions, by stymieing other kinds of investigation, was holding the field back. The same person just quoted also described, with no small passion, how neuroimaging had:

> really overshadowed experimental psychology—that is, the examination of
> the psychological mechanisms underpinning behavior. But the fact that the
> technology excited people so much. And there is a whole swathe of research
> published in the last ten to twelve to fifteen years, particularly the earlier

stuff that simply is, "Oh, that lights up when you show them that," and, you know, not very much else . . . there's a sense in which unless it's got some neural signature, this research, it isn't of any validity. That has to be wrong. It's scientifically wrong.

Again, I am not suggesting that there is anything very revelatory or shocking about this claim. That people expressed these views to me without great prompting is fair evidence, in fact, of their mostly uncontroversial nature. I *am* trying to argue that the consistency of these kinds of claims, in which this specific group of people, whose professional identity is wholly or partly invested in doing brain-imaging studies of autism, but who nonetheless frequently position this practice as either partial, or flawed, or misleading, or invalid, or maybe just inappropriate to studying things like autism in the first place—that the preponderance of these accounts finally adds up to something noteworthy. This thick patina of low expectation, suspicion, anxiety, and critique, which seems somehow inseparable from so many researchers' accounts of their own basic neuroimaging practice, needs some firmer theorization.

ENTANGLING EXPECTATIONS

The generative role of positive expectation seems clear enough (i.e., hopes and promises make it more likely that actors and resources will gather around a scientific project or object). But the role of *low* expectations among this group of neuroscientists seems less obvious—and recall my earlier insistence that these scientists continue to work, even while maintaining these low expectations, without at all conceding that a deflationary rhetoric requires them to stop gathering-together a neurobiological account of autism. *Tracing* has been my word to hold all of this together: it describes the difficult connecting, marking, and diagramming work of the neuroscientists I have interviewed. But it does without assuming that all that's happening here is an artificially human practice; the word clings onto an idea of *something being traced* all the same.

In the introduction I claimed some affinity for this usage with the "agential realism" of Karen Barad—a more thoroughly developed theoretical apparatus for thinking about the independence of nonhuman agencies, even while recognizing the fundamental entanglements and ambiguities of worldly phenomena. Of course, "agential realism" is one among a group of related terms and phrases that try to do similar work. Donna Haraway (1997: 268–69) has usefully provided a list, including modest witnesses, boundary objects, situated knowledges, and misplaced concrescences—to which we could probably also add "vibrant matters" (Bennett 2010), "arche-fossils" (Meillassoux 2008), "quasi-objects" (Serres 2007), and no doubt a whole host of others. While all of these carve out their own theoretical space, what they share is a desire to provincialize human interest and practice within the spaces of both ontology and agency. They share the recognition of a world of nonhuman objects and agencies that may well be caught up in (and partly generative of) human affairs, but that also may be sometimes entirely indifferent—and, indeed, actively indifferent—to human concerns with language, culture, meaning, symbol, and so on.

What attracts me to Barad's "agential realism" in particular is that it is deeply embedded in a refusal to separate the practice of science from the practice of studying science (i.e., from the outside): "the tradition in science studies," Barad (2007: 247) has pointed out, "is to position oneself at some remove, to reflect on the nature of scientific practice as a spectator." If we take the physicist Niels Bohr as indicative—for whom, Barad claims, "epistemological, ontological, and ethical considerations were part and parcel of his practice of science"—then we can begin to think about the ways that answers to the traditional questions of the social studies of science might *also* be intrinsic to the scientific practices in question (ibid.). This suggestion—that understanding the entangled complexity of scientific practice might mean giving up on a sharp distinction between "science" and "science studies"—has been vital for helping me to see the intricate ways that neuroscientists work through the larger epistemological and ontological fields in which their practice is implicated (Barad 2011: 446).[5]

At the heart of Barad's proposal is a focus on constitutive relationships between the mess and ambiguity of entanglement, and the strange possibility of distinction or singularity—with the latter coming after, not before.[6] In other words, the more common "interaction" assumes that distinct agencies come before their interaction with one another; for Barad, it's the other way around. What is at stake in the deployment of intra-action is that, for Barad

(2007: 136–37), an agential realism "does not take separateness to be an inherent feature of how the world is. But neither does it denigrate separateness as mere illusion . . . relations do not follow relata, but the other way around." Reversing (but not annihilating) the relationship between being apart and being together, Barad's metaphysics centers on (in her terms) "phenomena." Through this term, Barad (ibid.: 139–41) insists on holding together "the ontological inseparability/entanglement of intra-acting agencies," on the one hand, and the "primary ontological units" of the world, on the other.

This is to say, if relations come first, it doesn't follow that everything is relation. This is important. Distinction-making "agential cuts," in which distinctive "other" entities are sundered from the great mass of entangled agencies, are, in Barad's account of phenomena, quite capable of setting things apart from one another (ibid.: 333–34; cf. Bruno Latour [2012: 3] on "plasma" and on the variety of kinds of "explication" and "composition" that can come out of it). It is, in the end, through this sense of a cut *within* phenomena, that Barad (2007:183) rescues objectivity from entanglement, in which she sets herself apart from the "positioning of materiality as either a given or a mere effect of human agency." In other words, the ceaseless intertwinement of human and nonhuman, biological and social, does not leave us with a hopeless morass of entangled identity and sameness. As media and cultural theorist Lisa Blackman (2010: 171–72) has pointed out, dynamics of affective and psychic entanglement are not in opposition to the emergence of objects; movements and intertwinements may be more accurately characterized as forms of interiority and inhabitation from (and within) which objects, entities, and agencies are cut.

I am a long way now from neuroscientists talking about whether they are still hopeful about the prospects of brain imaging or more disappointed by it. But Barad's work forms a crucial part of the general hinterland of my way of understanding such talk—as a practice that seems committed both to the inseparability of its objects and practices, yet attached to the agency and independence of those objects all the same. I raise this here, particularly, because the intricacies of Barad's framework give some good reasons why we might not actually see a contradiction in the shifting dynamics of hope and disappointment within these interviews, at least as these map onto a similarly mobile relationship between entanglement and distinction or between the act of tracing and the thing that gets traced.

Autism neuroscience is characterized by a sense of sometimes frustrating entanglement, but it also remains committed to, and hopeful of, something *beyond* all the same. It is in this sense that a tracing neuroscience of autism is perhaps always going to be at least partly a marriage of hope and disappointment: these wavering views on the future of autism neuroscience closely echo the dynamics of entangled complexity and cut-away singularity that are inherent to just this kind of intra-active pursuit. So the shifting copresence of promise and disappointment begins to seem understandable—even necessary. My suggestion, at least, is that such dynamics should be heard as ways of talking about, and dealing with, the hard work of a tracing neuroscience—a work of connecting and diagramming and laboring; of seeing the inseparability of your own work from the thing worked on but still remaining committed to a cut all the same. It shows how and why a sense of disappointment might actually be *built into* an account of generative expectation in analyses of technoscientific projects. It shows the role that the sensation and expression of this more dynamic imaginary landscape might yet play in the generation and sustenance of scientific work.

CRITICAL NEUROSCIENTISTS

This runs against the grain of some recently prominent ways of thinking about the sociology of neuroscience. In the introduction I briefly discussed (and set myself against) self-consciously "critical" accounts of neuroscience. One of the most potent of these interventions has been made by a group of scholars arguing for what they call a "critical neuroscience" (Choudhury, Nagel, and Slaby 2009; Slaby and Choudhury 2011; Choudhury and Slaby 2011). The essence of the "critical neuroscience" argument is not to tear down neuroscience but instead to inculcate among neuroscientists "self-critical practices, which aim to achieve reflective awareness of the standpoint-specific biases and constraints that enter into the production, interpretive framing and subsequent application of neuroscientific knowledge" (Choudhury, Nagel, and Slaby 2009: 65).

As the cultural theorist Felicity Callard and I put it elsewhere (and I draw on and expand that discussion here), the claim is that "neuroscience *itself* should be reformed as a critical practice—insofar as it must become aware of its own political and economic (among other) standpoints and drivers" (Fitzgerald and Callard 2015: 10).[7] To that end, Suparna Choudhury and her colleagues in this tradition identify neuroscience, and the "brain facts" it produces, as a practice beset by the social and economic context in which it finds itself, as something caught up with economic drivers, political climates, and cultural contexts (Choudhury, Nagel, and Slaby 2009: 65). "Brain facts," they point out, are not "objectively given things-in-themselves but emerge from communities of scientists working collectively at a given time in a given context" (ibid.). The argument is not that brain facts (on the loose) *become* cultural products, but that they *are* cultural productions, by definition, and that neuroscientists need to begin taking account of this.

One of the main drivers of this intervention is an attempt to question the attention to neuroscientific accounts more widely. As my colleagues and I put it elsewhere (Fitzgerald et al. 2014: 2), the goal

> is to question the broader cultural urge towards neuroscientific explanations, to point to the problematic bases both of this urge *and* of the brain science it wills into existence, and to imagine, beyond both, a different sort of neuroscience—one that is able to question its own "givens" and to recognize its own history and context; a discipline in which "historical, anthropological, philosophical, and sociological analysis can feed back and provide creative potential for experimental research in the laboratory." (Slaby and Choudhury 2012: 29–30)

Particularly important is what Slaby and Choudhury (2011: 31), in their "Proposal for a Critical Neuroscience" have advocated as a "political" rethinking of neuroscience—that is, a concern with the "cultural tendency favouring voluntarist conceptions of the self" and an awareness of the "correspondence between economic imperatives and normative schemes, and so on." At the heart of this claim is a desire to see both neuroscience, and the things that neuroscientists work on, as fundamentally enveloped in a social context: "the brain and nervous system are nested in the body and environment from the outset . . . their functions can *only* be understood in terms of the social and cultural environment" (ibid.: 33, my emphasis).

To the extent that such "social factors" are allowed into neuroscience, this is only to the extent that, as the historian Nancy D. Campbell (2010: 91) has pointed out, they remain "reductive and abstract, rather than concrete and substantive." The concern, then, is that these kinds of things will either be elided or irredeemably reduced by a neurobiological imperialism. As Choudhury and her colleagues have remarked: "While psychological distress no doubt has manifestations at the level of the brain, the biological claims free the person from the social and cultural complexities surrounding her . . . future advancements in neuroscience will ensure the displacement of several psychiatric practices including psychodynamic, social and cultural psychiatry—by biological approaches" (Choudhury, Nagel, and Slaby 2009: 71). Let me make a very simple observation. While undoubtedly there are public spaces of neuroscientific discourse that could use some more self-awareness, the data discussed here, on the entangled expectations of my interviewees—and of the tracing neuroscience that these expectations bespeak—must remind us, all the same, that sometimes neuroscientists are in fact strikingly critical of their own practices.

More to the point, they frequently express this critique, and their wider sense of deflation, precisely by recognizing the implication of their science with some wider social context. For example, in a discussion of whether or not brain-imaging studies were actually very reliable, a lecturer argued that one of the main problems with neuroscientific autism research is that

> a lot of research doesn't work with those [more "impaired"] people because they're just so difficult—certainly an adult—to work with, because they don't understand, they're frightened, they can't cooperate.

Or as another interviewee put it even more strongly:

> The world is not as tightly controlled as in an RCT [randomized control trial] trial. . . . In RCTs, you chuck out all the people you don't want—so, you know, chaotic families, parents who don't speak English, parents who are a bit bolshie, kids who are a bit bolshie, and you've usually got particular criteria for IQ and language . . . so all the difficult kids and difficult families aren't involved.

In one memorable anecdote, an ex-employee of a funding agency told me about a conference that her organization had been involved in, which brought together scientists involved in different kinds of cutting-edge neurobiological research on autism. But there was a twist: "We arranged as a final session," she said, "for a member of our staff to present who has a son with autism:

> And she just described sort of, you know, how it had been, and what she'd done, and what she'd tried, and for the first time, she talked about—he has a number of compulsive behaviors and things—and she talked about what these fixations and compulsive behaviors were in practical terms, and the implications of them. And it was transformative for a number of researchers in the room. There were researchers in tears—she wasn't; but they were. Because I think this was the first time for some of them, particularly those who were working on mouse models and things like that, repetitive behaviors in mice are not the same thing as repetitive behaviors and compulsions in humans. And she just gave some practical examples [. . .] and a lot of people went away and really rethought, you know, what they were actually in the business of doing—and I think started to see for the first time that they were actually in the business of trying to reach therapies that improved the outcomes for people with autism.

Here, the question of therapy, the question of neurobiology, the question of what people like my interviewees "were actually in the business of doing," are all deeply entangled in the jolt of one person's confrontation with the daily life of another. Notice how tearfulness and the confrontation with realities of social life are not narrated here as something inimical to science. I hear this not only as an expression of the kinds of low expectations that I have been trying to call attention to throughout the chapter ("repetitive behaviors in mice are not the same thing") but also exactly the sort of critical sensibility that Choudhury and Slaby and Nagel are seeking—namely, a recognition of entanglement and context, a rejection of any kind of "naïve" objectivity, and the encouragment of a much more modest set of clinical and therapeutic hopes.

Focusing on the complex interplay of expectation and disappointment gives us a glimpse of how neuroscientific discourse is not only much more complex and sophisticated than its critics allow it. That discourse often contains a much more entangled view of the relationship between the biological and the social, or the human and the nonhuman, than sometimes the critics *themselves* actually hold. It shows us not only how tracing neuroscience is a more ambiguous practice than much sociology and anthropology of science recognizes. It shows a much deeper and richer form of entanglement than most sociological and anthropological accounts are themselves capable of. Hope and disappointment rub against one another here, not only because an "objective" neuroscience is always embedded in some "social" context but because the mutual entanglement of these two, and the complexity of the relays between them, requires some very nimble intellectual work.

So what of *critique*? Of course I am being simplistic in how I wield this word—there is obviously a complex, variegated literature on the history and present of the critical impulse, as well as the nuances of how it might (or could) be mobilized in and around science, that I am riding over in a fairly rough way (see, e.g., Folkers 2016). Still, my usage might not be so much less complex than the loose way that "being critical" gets unreflexively valorized in the social sciences, especially in any kind of encounter with biology. I have set out elsewhere, with different groups of colleagues, what it is I find so tedious about this impulse (Fitzgerald et al. 2014; Fitzgerald and Callard 2015) and what kind of approach I think should take its place (Fitzgerald, Rose, and Singh 2016). What I draw attention to here is the sheer *redundancy* of the critical theorist's wagging finger: not only are these scientists quite capable of contextualizing and problematizing the different intellectual and political territories of their own practice, but they are able to fold these perspectives back into the mundane daily labor of scientific practice, to produce a much richer, much less settled, and much more entangled account of the crossing of bodies, affects, and politics. Whatever it is that social scientists have to contribute to what we know about the neurosciences, it cannot be anything that goes under the sign of *critique*.

COMPLICATIONS

Starting to emerge, of course, is one of the more obvious subterranean themes of my suggestion that the complexity and ambiguity of the new brain sciences has not always been well appreciated by an interested social science. I am positioning this book within a particular way of thinking sociologically about the natural sciences, one that refuses to take the "social" and the "natural" to be different kinds of thing, or to let one explain or override the other—remaining attuned instead to the constant coproduction, entanglement, mingling, and colocation of these categories (Haraway 1985; Latour 1993). In the introduction I suggested a rough division within social science accounts of the new brain sciences. I described a literature that grants social interaction priority over the operation of technologies, bodies, and nonhuman agencies, and which creates a hierarchy of explanation on this basis (Martin 2000, 2004; Ortega and Vidal 2007). But I also described a literature—I used the word *reparative*—inclined to see a more symmetrical interplay between these categories, one that sees the new brain sciences as an area both structured by, and generative of, this symmetry, that therefore maintains a trickier (but no less rigorous) relationship to methods of explanation and description (Wilson 2004; Rose and Abi-Rached 2013).

What emerges, then, is one of the more significant undercurrents to the book. In aligning my account of tracing neuroscience in this chapter with Karen Barad (2007), but also, in upcoming chapters, with a range of other feminist, cultural and social theorists, as well as the occasional philosopher, including Elizabeth Wilson (2004), Alfred North Whitehead (1979), Bruno Latour (1987), and Donna Haraway (1997), I am stringing together—perhaps erratically—a nexus of affiliation and disaffiliation. I position the book within a web of scholarship that is marked by an attention to the seriousness of nonhuman material, marking a commitment to seeing that material as ontologically and agentially symmetrical to human categories. I return to this discussion in more depth in chapter 5 and in the conclusion.

3

THE THROBBING EMOTION
OF THE PAST

Memory, Scientific Subjectivity, and the Affective Labor of Autism Neuroscience

PROBABLY THE MOST SURPRISING THING FOR ME, OUT OF MY MANY hours of conversation with autism neuroscientists, was how emotional our exchanges sometimes got. It didn't always happen, and when it did it wasn't always very potent. But it remains a cold empirical fact that neuroscientists, especially those who work on autism, sometimes talk about their research in strikingly emotional narratives. It is equally a fact that they sometimes trace their scientific work through the bodies and feelings with which such emotions are registered. Of course, emotion has long been at stake in autism research (Baron-Cohen 1991), but my focus is not on the emotional lives of people diagnosed with autism. I am interested, instead, in flows of emotion and feeling within the bodies of autism neuroscientists themselves: as much as neuroscience might be an identifiably *intellectual* or *technical* endeavor, the neurobiological account of autism is also traced through some unambiguously affective and emotional labors.

This was actually not a planned interest. I set out to do the interviews for this book with a self-conscious commitment to (what I imagined to be) *conceptual* discussions about the way that autism neuroscience was done. But for the interviewer not so well endowed with confidence about the field he or she is entering (i.e., me), there is no clear entry point to talking *conceptually* about how a solid neurological account of autism might come about. Nor is there an obvious intermediate discourse that would link such a question with the interests and concerns of (mostly) neuroscientific interviewees. Quite early on, and purely as a device to get conversations going, I started asking scientists about their initial motivations for doing neurobiological autism research. I asked them to tell me the story—if there *was* a story— of their entry into autism research, the things that grabbed their attention within it, and the questions, opportunities, or concerns that pushed them forward. I expected in reply a litany of books read, lectures attended, mentors cultivated, and intellectual interests developed. Sometimes that's what I got. But other times, I got stories that were very different, stories whose major themes were not guiding theories, powerful explanatory paradigms, or key figures. These were stories about the quality and depth of the feelings experienced by the individual scientist in the course of his or her research. They were stories of upset, sadness, and fear but also stories of pride, desire, and even love.

MOTIVATIONS

In her account of the emergence of the autism spectrum and of the waxing understandings that have appended it, STS scholar Chloe Silverman has urged attention to the use of love, specifically, as an "analytical tool" for the social study of biomedicine. Talking about love, writes Silverman (2011: 3), "shifts the focus from psychiatrists, epidemiologists, and geneticists to parents, counsellors, diagnosticians, and lawyers." If, for psychologists, love has simply been something that they studied in autism, nonetheless "beyond the laboratories . . . love continues to function in normative claims about the practice of research. Parents and their allies say that emotional knowledge enables them to observe and attend to their children in the right way, guides them in medical decisions, and helps them make the right choices for the person they love" (ibid.). For Silverman, if love has sometimes been seen "as a liability or a barrier to reliable knowledge," there is room now to start thinking about love as "the source of specific, focused and committed knowledge" (ibid.: 3–4). Throughout this chapter I advance Silverman's suggestion in two ways. First, I build on her account of love, to think about the broader role of emotion, and of affect more generally, in the putting-together of serious, concrete knowledge. Second, though, I argue that paying attention to emotions, and to the role that emotions play in assembling an account of something like autism, does not at all move us beyond the laboratory. In fact, focusing on the role of emotions in making knowledgeable claims tells us something important about scientific work: here, and among autism researchers especially, I show how an unimpeachably scientific, laboratory-based work of looking for and thinking about the neurobiology of autism is often an emotional and an affective labor too.

Of course, the recognition of emotion's presence in scientific spaces has a long history in the social study of biomedicine (see, e.g., Lynch 1985; Latour 1996). But if the place of affect in the laboratory is not a surprise, there still remains a question—most famously posed by Max Weber (1919)—about the relationships between the passionate attachment that motivates a scientific interest, the actual performance of scientific work, and the status of the objects and truth claims that emerge *from* that work. Recently, and drawing in particular on long-standing attentions to the body and affect within feminist science studies (Haraway 1988; Hayles 1999), scholars have started to work through the entangled nature of this relationship. The feminist theorist Elizabeth Wilson (2010: 24), for example, in her work on Turing's calculating machine, has called attention to the way that "feeling and thinking might coassemble" in the unfolding of a modern technoscience. Anthropologist and STS scholar Natasha Myers (2012: 153–56), focusing on the relationship between dance, body movement, and molecular biology, has shown some of the ways that scientists' body practices can become "effective media for articulating the forms, forces and energetics of molecular worlds." This underexplored mixture of body, feeling, emotion, and science is what I, too, am working to understand. But I also want to know whether the stories I relate here can open up an account of the neuroscientific life more generally. How might these memories, and the emotional weight attached to them, help us to think through the affective hinterland of high-tech bioscience and the people and objects that emerge within it?

In the introduction to this book I said that creating neurobiological accounts of complex diagnoses is a more ambiguous and intricate practice than is yet realized in much of the literature: *tracing* is my image for talking about the ways in which autism neuroscientists seem to live with, and work through, this ambiguity, while not abandoning their commitment to some objective neurological account of autism all the same. I expand that discussion in this chapter by thinking through the copresence of an affective laboratory labor on the one hand and the desire for some kind of biomedical concrescence on the other, drawing on the process philosophy of A. N.

Whitehead (1935, 1964, 1979). This is not as esoteric a leap as it maybe looks: Whitehead was committed to thinking seriously about the way things come into the world, and how they are held together in it. But he also offered an analysis of these processes as specifically *emotional* moments of connection and assemblage. Thinking with emotion, for Whitehead, was also a way of thinking about how things come to exist. As I show through these accounts of emotional memory, paying attention to how Whitehead held such ideas together—emotion, intersubjectivity, concrete existence—might be surprisingly important for understanding the contemporary neurosciences.

"I REMEMBER ONE WOMAN. IT WAS HEARTBREAKING"

Let me leap straight in to the stories at the heart of this chapter. I begin with an autobiographical memory that comes from an interview I did with a professor of cognitive neuroscience. This was an early interview, conducted when I had yet to solidify my approach. The extracts I focus on here, appearing unbidden in the middle of a conversation about this professor's early career, actually first taught me to listen for the affective commitments that might circulate in scientists' accounts of their own intellectual work. We were talking, in a fairly loose way, about her research before obtaining her PhD. This professor began to tell me about her first real job in the field, which involved traveling to schools around the United Kingdom in the 1980s, doing psychological tests on children with diagnoses of autism. She began:

> I was traveling around on my own. And I remember the first time I walked into an autism school—which I think was [*names a well-known school*], one of the first autism schools there was. This of course was in the late eighties. And . . . it's hard to convey because actually autism schools aren't like that anymore. But . . . the sound when you go through the door—the kind of particular sounds that low-functioning children with autism make in place of language that aren't like the squeals of ordinary kids. To just walk into the playground and see all the children busy, all this movement but none of it coordinated, children not moving together—you know, one child in a corner rocking, one child examining the light through their fingers, another child running in circles, with this extraordinary squeal that other

children can't produce. It's just so different. You walk through a door, and
on the other side of the door is a world that's so utterly different from the
high street in [that part of London]. So I found it completely captivating,
and . . . terribly, terribly upsetting.

"That break between the ordinary world and the world of the school," I
asked, "can you remember . . . on what level did it hit you? Was it kind of an
emotional, kind of . . . ?"

It's very visceral, yeah. Very visceral. And although I thought I knew a
lot about autism, because I'd *read* a lot about autism, I'd *heard* a lot about
autism—actually I was utterly unprepared for it. Nothing I'd read conveyed
the level of lack of language, intellectual impairment (although of course
that may be secondary to other things but, you know, presenting intellectual
impairment). So, here was this thing that I was really passionate about and
interested in intellectually. And then it hit me in the stomach.

It's maybe an elementary point, but it is important to note that this account,
as a description of entering the field and actually doing science, is already in
a markedly different register than those proposed by most formative sociolo-
gies and histories of science. Robert Merton (1979: 270), the pioneering sociologist of science, famously described the "ethos of science" as a structure
made up of "universalism, common ownership, disinterestedness and organized scepticism." For Merton (ibid.: 259), the "sentiments embodied in the
ethos of science" would best "be characterized by such terms as intellectual
honesty, integrity, organized scepticism, disinterestedness [and] impersonality." But what is notably lacking in Merton's description is what is actually
primary in the extract above—that is, the memory of a specifically emotional and affective hinterland, behind this (still, insistently) scientific ethos.
This isn't just an aspect of scientific practice that is left unaccounted for by
Merton—in fact, it is very clearly anomalous to his "ethos" as such. Note,
for example, the degree to which the memory is made up of such images as
"upset," "viscera," "passion," even being "hit . . . in the stomach"—and contrast with Merton's description of the "disinterested zeal" of modern science,
in which a "puritan spur" allows "the exaltation of the faculty of reason" to
emerge as "a curbing device for the passions" (ibid.: 228, 238).

Obviously, it is true that Merton's sociology of science is no longer very influential (and also, however one-dimensional his view looks now, it cannot be separated from Merton's memory of the very *interested* scientific zeal that manifested during World War II). Nonetheless, similar themes of impersonality, disinterest, and distance still structure many accounts of what science is like—even where it is confessed that such tropes are only fit for public consumption. Consider psychology, for example, which, as Theodore Porter (1996: 211) has pointed out, following the historian Mitchell Ash (1992), "has been more self-consciously scientific than the natural sciences precisely because of its institutional weakness and intellectual disunity. Inflexible methods of quantification compensated for the lack of a secure community . . . [statistical tests] were part of a regime of replication and impersonality, necessary if the study of psychical phenomena was to win even a modest degree of scientific credibility."

But contrast *that* view of the disinterested and disembodied self-consciousness of psychology with the following memory—which came from a different, much more junior researcher, who also, to my surprise, related the ways that you inevitably "bring yourself" to the research process as a neuroscientist of autism. She said:

> I remember one woman. It was heartbreaking. Her child was very high-functioning, won awards for this, that, and the other. It was her only child—she said, "Oh, he was my blessing." She didn't think she'd have children—she was, like, forty-plus. But she said, "Oh, it's so sad, I'm there with my husband eating breakfast and just wish that . . . Daniel or whatever . . . would just sit and eat breakfast with us. But he's just not interested. He just gets up and walks away." And she was really grieving, because this relationship with this blessed child that she'd longed for just hadn't materialized, but he was incredibly articulate, incredibly bright, very nice boy—but that kind of emotional, just not engaged. I thought, "It's so interesting" because she's, you'd think she had the easier deal [i.e., than parents of "low-functioning" children with autism], but she was the one who I think was emotionally finding it a lot harder than some of these other parents, who you think, "Well, I don't know how . . . your lives are so disrupted." You know, so that was a really interesting lesson. I don't know how it informs my . . . but I think it does make me sensitive, and one thing it does make me think about is that, kind of issues

that people will probably say are to do with political correctness, I'm very careful about the language I use, like, you know, "individuals with autism," and people say "autistic." You're just, you know, very careful with the language. And being respectful, and things like that. And think, you know, these aren't, you know, your subjects—well, they are but, you know, you don't, that's not the way you . . . relate to them.

Here is the psychologist nether as disinterested puritan nor as impersonal seeker of scientific credibility. Instead, here is the psychologist as an emotionally invested, slightly heartbroken young woman—a self-consciously sensitive scientist, keen to stress her hearing of a mother's grief, fumbling a little with the politics of language, and worrying about whether or not she relates well to her *subjects*—a term that she is even reluctant to use. What was particularly striking about this account was that the researcher described the laboratory not only as a space laden with emotion, sensitivity, and heartbreak; in fact, she characterized her *scientific* work as a much more emotionally potent space than her life outside. "I think I'm quite an empathic person," she said later on. "I don't want to take this sort of stuff home with me."

Let me add a caveat before moving on. If affect is surprisingly present in this space, it is less of a surprise to learn that this affect tends toward the negative. That negativity reflects a sense of autism as defined by deficit and loss, an affective economy that in turn seems rooted in clinical and bioscientific understandings of what is at stake in autistic life—namely, a problem causing heartache (and not, for example, a difference provoking joy).[1] This opens up two overlapping debates: one of these, long-running and broad, is about the differently *social* and *medical* models of disability, and whether something like autism marks a "deficit" in an individual, as opposed to a more widespread inability to accommodate difference. A related debate, more recent and more specific, is about neurodiversity. This debate concerns distinctions between a medical or bioscientific account of autism as a problem to be fixed versus a perspective that takes autism as a set of differences, skills, or idiosyncrasies to be valued in their own right (see Robertson and Ne'eman 2008). Bear with me for a moment, as I set this important issue to one side. I do want readers to have it in the back of their heads as I move on though, and I'll return to this issue in the conclusion.

Here are two more stories about the absence of emotional distance in neuroscientific work on autism in which disinterest is not just elided, but rather the moral imperative of scientific work is specifically addressed and emotively affirmed. The first comes from a psychiatric neuroimager whose background was in physics, and who was then working in a major neuro-imaging center. He started out by talking about his (not wealthy) childhood in provincial England and about how his life and career had developed sub-sequently and thus his desire to make some kind of "contribution." In the middle of this conversation, he said:

> I was joking the other day with a friend of mine who used to work in the motor trade. We were talking about service managers [i.e., people who work in administrative and managerial jobs, as opposed to, for example, clinical roles]. And I'd just met some service managers who were friends of friends of ours, at a dinner party—and I was surprised at the lack of conscience that they had. They had no conscience. You know, their family and their friends were very important to them. But essentially if it didn't influence them, or didn't have an influence on them directly, they had very little interest in it. And it just seemed to match with the kinds of work that they had to do—you know, they were very keen on meeting business targets, and that's the priority rather than . . . people. And the way people feel. [. . .] You know, that's the way some people are and that's fine. But some people are the other way around, and they care perhaps a little bit more about the way people feel.

Of course, when he's talking about the "some people" who are "the other way round," he means himself and his colleagues—that is, bright, highly educated people, working to alleviate the problems of people with neuro-developmental problems and doing so in a fairly unglamorous office, for relatively small salaries. I was particularly interested in the way he narrated his relationship to scientific work as a way of attending to "the way people feel." There is a clear sense of the affective commitment of his science and of his self-construal as a researcher—indeed, as a physicist—precisely through registers of feeling and care. There is also an interesting feeling of almost moralistic pride running through his account, which may or may not be misplaced (is it really so reprehensible to worry a bit about business targets?), but which also again disrupts assumptions about the disinterested, impersonal nature of scientific work.

The same researcher told me a story about going with a group of friends to a comedy club some years previously, where he was picked out of the crowd by the comedian. When the comedian asked him what he did for a living, this scientist was pleased to be able to give the joke-defying response of "cancer research" (which he had worked on early in his career as a medical imager). "It's just about getting some sleep at night," he said to me, "so, I suppose if I can go to sleep at night thinking I've made some sort of contribution . . . then I'm happy." This idea, of worrying about being happy in scientific work and in locating that happiness in a commitment to care for other people, was echoed by another postdoctoral researcher. She initially maintained a distance from any kind of passionate or emotive account (I was by this stage on the lookout for them), telling me that she had gotten into autism "by chance" and that her early interests were in mathematics and neuropsychology particularly.

Then the researcher went on to talk about how important it was for her to work directly with parents of children with autism, and with their teachers. She described the qualities of those interactions, and their centrality to her work, in a way that I thought was very obviously heartfelt and that belied her earlier account of simple intellectual happenstance. I asked her about the difference between the two ways of accounting for her early interest in autism. She said:

> For myself, personally, if I was doing more of a pure science that wasn't . . . didn't have an end-point that was kind of applicable—I'd sort of find that quite difficult. I think I need to, um . . . either be a clinician myself [*laughs*] or do applied research.

When I asked her why she couldn't just be happy producing data, she said:

> Um . . . [*long pause*]. I suppose it's just that satisfaction . . . that whole knowing that what you're doing has a purpose. Um . . . and that it's not just a job. You're dealing with people's lives and, you know, a report that you write about a child does go on and have an impact. So, um, it's . . . yeah [. . .].
> Well, it's . . . the fact that, you know, you're listening to parents and hearing their stories and you're, you know, hearing their suffering and what they've

had to put up with and the fights and battles. And, I suppose . . . the fact
that they give a lot to you in taking part in research. So they are, um . . . it's
very different from a clinician's role. So they give a lot of time to you. And
so I suppose it's just wanting to give something back to them.

She then talked more generally about talks that she gives to parents and
clinicians, and how she contributes to the field more generally, before laughing self-consciously and saying, I think only half ironically: "I hope I have
a purpose in life."

What I find striking, in these last two accounts, is the sheer refusal of
disinterest. I am working to understand an unexpected sense that scientific
work is privileged in how it attends to other people's feelings—an awareness
that was, both of these scientists emphasized, *really* important for them. In
this chapter I am trying to account for the way that these researchers trace
an objective and scientific neurobiology of autism precisely through such
affective registers. I am interested in how an "intellectual interest" in the
neuroscience of autism runs parallel to accounts of stomachs, feelings, suffering, and heartbreak.

I WANT TO DO THIS. I WANT TO DO THIS

In their history of scientific objectivity, Lorraine Daston and Peter Galison
(2007: 203) have pointed to the emergence of a new and rather particular
kind of scientific figure in the nineteenth century: whereas the "enlighten-
ment savant" had been an active and critical synthesizer of data, the scientific subject of the nineteenth century "strove for a self-denying passivity."
Far from the almost "otherworldly" figure of previous eras, the goal now was
to "practice self-discipline, self-restraint, self-abnegation, self-annihilation,
and a multitude of other techniques of self-imposed selflessness" (ibid.). In
his account of the scientific life, similarly, the historian of science Steven
Shapin (2010) has identified a turning point in the past couple of centuries,
in which the passionate, feeling scientist was slowly effaced by narratives
centered on slow, processual, collective endeavor. For Shapin (2010: 33), the
key moment was a shift in emphasis to "method"—and the eclipse thereby
of the lone genius. Thus, Shapin argues, did we move from the unashamed

passion of Benjamin Franklin in the 1770s to a "stress on mundane method-ological discipline" a century later—and thus from speculation to technique, from metaphysics to facthood (ibid.).

Scientific work had moved from being the knowledge achieved by an indi-vidual, feeling body to being the faceless abstraction of a collective process. Strikingly, Daston and Galison (2007: 216) record that whereas eighteenth-century hagiographers of Isaac Newton knew him as the font of "by far the greatest and most ingenious discovery in the history of human inventiveness," for their Victorian counterparts he was something very different—namely, the epitome of "self-control in speculation, and . . . great-souled patience in the pursuit of truth." In our century, writes Shapin (2010: 47), this has devel-oped into a claim for the "moral equivalence" of scientific practice—a drive, particularly in the wake of the technological horrors of the twentieth century, to stress the moral ordinariness of scientific labor and its consequent abstrac-tion from the vices, virtues, and feelings of individual scientists.

Of course, contemporary scientific practices need to be carefully distin-guished from the establishment of these historical norms. Not least, the dis-persed laboratory of a modern technoscience (and the spread of this book bears witness to this) is filled with all kinds of bodies that were perhaps not well anticipated by the methodical and disinterested men of the eighteenth century—including, as feminist theorist and STS scholar Donna Haraway (1997: 269) has reminded us, women, people from the working classes, people of color, and many other inhabitants of "nonstandard" corporality and affect. Certainly, norms of disembodied self-abnegation still have some valence. One of my interviewees once told me about a problem she had when she used the word *thrilling* to describe her own attitude in a journal special issue and how objectionable a reviewer found this usage: "I really don't think this is the way to write scientific papers," was the judgment that came back. Yet her attempt to use the word in the first place, as well as her unruffled amuse-ment at the memory of the priggish reviewer, bespeaks change. Whatever public norms have emerged and receded in the past couple of centuries, it is unlikely that contemporary sciences track their ebb and flow with great fidelity. As Shapin reminds us, there is a danger of mistaking the public script of scientific propriety with the emergence and entanglement of actual technoscientific work: the fact is, writes Shapin (2010: 5), "the closer you get to the heart of technoscience, and the closer you get to scenes in which technoscientific futures are made, the greater is the acknowledged role of the personal, the familiar, and even the charismatic."

I am trying to get at this shift away from the analysis of norms of disembodiment and method, and toward thinking about the role of the individual and the familiar "at the heart" of technoscientific "scenes"—categories that I read in the distinctively embodied and emotional stories of scientific labor related earlier in the chapter. I am particularly thinking about the way that a contemporary scientific project, even one as dryly statistical and methodological as the search for a neurobiology of autism, is traced through the feeling body of an unapologetically individual, familiar neuroscientist. In addition to the accounts of emotional laboratory work just described, the role of emotions and the body were evident in interviewees' recollections of the formation of their scientific subjectivities as such, including their affectively registered *desire* for the scientific life—both memorialized as strikingly revelatory and emotional moments. One interviewee, remembering her decision to become a psychologist, recalled:

> When I was thinking about university, I'd really had a very—I was intending to do English literature, actually, and then just one day, literally, I woke up and thought, "No, I want to change that to psychology," having no really formal idea about psychology education. [. . .] Looking back on it, I still think that that was how it was. It was literally a wakeup in the morning and say, "I'm not going to do English literature. I can still read the books I want to read. I'm going to do psychology."

She was not alone in embedding the formation of her scientific subjectivity in a moment that was entirely personal, passionate, and embodied. The professor whose memories opened this chapter also mentioned an early encounter with autism parent Clara Claiborne Park's (1982 [1967]) pioneering memoir as something that "made me very excited." One graduate student even related how, when still at school:

> I was quite into literature and drama, and quite creative, I suppose—but I couldn't like really figure out ever, "How am I going to get a proper job out of this?" [*laughs*] And then we were doing Dr. Jekyll and Mr. Hyde at school, and one of my teachers recommended that I go and read some Freud because this

ties in, so then I went to the library and I looked up Freud. And I remember—
I just specifically remember—being, like, people research how other people
think and feel? And I never knew this before [. . .] this was a massive revelation to me, and I was like, "I want to do this. I want to do this."

Max Weber (1919) famously described the scientist's feeling of "strange intox-
ication," which, along with a "passionate devotion," Weber was very well
aware of, even if he was very keen to rigorously separate interest from practice in the scientific vocation. But if, for someone like Weber, a passionate
zeal must remain distinct from the labor of science, I am trying to show
how contemporary scientists sometimes work across those boundaries in a
much more complex manner. During these interviews, over and over again,
I heard accounts of scientific subjectivity and of scientific labor that bespoke
a much more entangled relationship between intellectual and affective
practices. In these interviews, doing the neuroscience of autism emerges
as the product of an intoxicating revelation, a revelation borne out in the
emotional and affective commitments of a very particular kind of scientific
subject—here, someone who feels, desires, and imagines; someone who is
easily thrilled but also slightly upset; someone who hears other people's
suffering; someone who is morally good but also someone who needs to feel
happy about that goodness.

THINKING AND FEELING

Over the past fifteen years or so, *emotion* and *affect* have emerged as major
points of inquiry in social-scientific and literary theorizing, to the extent that
some have even diagnosed an "affective turn" in these literatures (Clough
and Halley 2007; Gregg and Siegworth 2010). I am not interested in seeking
affinities with this development.[2] I am especially uninterested in parsing
the analytic specificity of the emotional or the affective per se (the distinction between the two, which I do not follow, belongs to Massumi 2002).
What I *am* interested in is the broadly affective, emotional, and embodied
way that some autism neuroscientists talk about their scientific work and
their scientific lives. It just seems so much at odds with the idea of the coldly

reductive neurobiological imperialist, marching unblinking toward her epistemological destiny—at least as we have come to imagine the prototypical neuroscientist in the social sciences. Obviously I caricature here. But still, what happened to our single-minded, all-conquering neuroreducer? Who is this excitable, sentimental figure I have found in her stead?

When I talk about the place of *emotion* within neuroscience, I recognize the broad nature of this term. While I have no great objection to the careful and parsimonious theorizing that surrounds the concept elsewhere, such an approach does not especially fit the empathic and "hospitable" method demanded by the jostling, circulatory nature of my data (Wilson 2010: xi). Indeed, if this chapter is to claim any affinities with the broad literature on feeling, it comes not from affect studies per se but through more long-standing material-feminist and queer attentions to the question of the body—a corpus that, for some time now, albeit without necessarily setting out from the study of the emotions, has been looking for ways to think and live with the always hybrid nature of corporeal human life (see, e.g., Haraway 1990; Grosz 1994; even Butler 1993 to some degree; for overviews, see Alaimo and Hekman 2008, or Hird and Roberts 2011).

Within the texts on emotion that are broadly affiliated to this tradition, some have recently begun to focus on the role of bodies and emotions within scientific and technological spaces especially. In a discussion of the life of Alan Turing in *Affect and Artificial Intelligence*, Elizabeth Wilson (2010) has explored the role of affect not only within Turing's own life and practice but within artificial intelligence and robotics more generally. At the heart of Wilson's discussion is the attribution of affective registers not only to the human scientists but also to the nonhumans in these spaces: machines, robots, and programs appear not just as figures or elements of emotional exchange but as much more complicated and active relays, both figured and figuring within a broader affective entanglement. Wilson is keen to stress the degree to which machines and machine properties (her focus is the automatic calculating machine of early artificial intelligence research) are not exemplars nor are they *products* of a cold, knowing intelligence. Rather, machines as well their creators are bound up with "a fusion of intellect and muscle and beauty and nerves" (ibid.: 8). Within the laboratory, Wilson argues, the relationship of thinking to feeling is one of "introjection" (an integration or a bringing inside), which significantly refigures how we think about the interiority of both the rational scientist and the calculating machine as well as the creative and generative back-and-forth between the two.

This more generous definition and attribution of affect opens a door for thinking about the role of emotional salience in technological and scientific labor: if Turing's work and life are animated by both "affective and intellectual concerns," then "it is his errant curiosity, his capacity for enjoyment and surprise, and his childish engagement with computational machinery that underwrite the importance of [his canonical] 1950 paper" (ibid.: 35). Wilson's account provides a picture of an affectively committed scientific and intellectual labor, in which the scientist is not simply beset by feelings for her objects but is instead one node within a much broader circulation of "affective commerce" (ibid.: 16; cf. Ahmed 2004 on the generative role of "affective economies"). The scientist's ability to recognize and negotiate the complexity of this emotional web is intimately entangled with her scientific practice. In Wilson's (2010: 16–18) account, Turing appears as a brilliant scientist not least because "the traffic between [his] internal states and the internal states of others is a key methodological concern in his work."

In a similar way, Natasha Myers (2012: 153) has drawn on her experience with the "Dance your PhD" phenomenon, and on her ethnographic work among microbiologists, to show some of the ways that scientists use their bodies and body movement to "figure out" the subjects of their research. Dancing and bodywork, Myers (ibid.: 153–56) has argued, can become "effective media for articulating the forms, forces and energetics of molecular worlds." Myers, like Wilson, calls attention to the "affective entanglements" of scientific research—or the way that the sometimes self-described cognitive or intellectual aspects of scientific labor, as well as the objects of that effort, are not easily separable from the feelings, the movements, and the bodily vibrations of the scientists in the laboratory. Using the body "can generate both new forms of knowing, and the things known," Myers (ibid.: 162, 171) points out; it can also "make explicit the kinaesthetic and affective dimensions of what are normally recognized as thought experiments."

While Myers (ibid.: 177, emphasis in the original) is particularly focused on dance, and on the cognitive-affective work of bodies-in-motion, she also calls for a larger attention to "researchers' capacities to *move with and be moved by* the phenomena that they attempt to draw into view." Bodies are "*excitable tissues,"* she argues; they have "the capacity to collect up and relay nuanced molecular affects" (ibid., emphasis in the original). One of the primary metaphors that Myers draws on in her account of scientists' bodywork is *rendering*— a term whose multiple valences carry the sense of a representation, of the work of producing that representation, of a cut, and also of a communication. There are strong affinities between what Myers has described as *rendering* among the molecular biologists in her study and the work that I have called *tracing* among the neuroscientists in my interviews. In both cases, what the term is trying to nuance is the awkward relationship between the production of something and the independence of the thing produced. "What holds all of th[e] uses of the term [rendering] together," Myers (ibid.: 172) argues, "is that each refers not just to the object that is rendered, but also to the subject, the one who renders, and the activity of rendering." "Dancing your PhD," like feeling your robot—like *tracing* your neuroscience—is an act of rendering: it articulates the entanglement between the body of the researcher and the thing researched. But it does so precisely in the service of animating, collaborating, *dancing* with the object in question.

With Myers and Wilson we begin to see not only how the intellectual work of the laboratory might be a very emotional and embodied experience. More specifically, we see the productive and even generative role of affect in scientific work. In Wilson's account (quoting Mike Fortun [2008]), "cognition and affect 'feed off each other . . . and set possibilities in motion'" (Wilson 2010: 22). For Myers (2012: 156, emphasis in the original), "scientists . . . conduct *body experiments* to work through hypotheses about how molecules interact." There are important differences between these two accounts—but, for now at least, I draw on what unites them, which is a suggestion that high-quality intellectual work, within scientific and technological spaces, is often deeply embedded in the ability of scientists to give themselves up to a kind of affective relationality. The capacity for thought is thickly entangled in the desire to feel.

HOW COMPLETELY FASCINATING, AND HOW COMPLETELY TRAGIC

Let me return to the senior professor I quoted at the start of this chapter. Having told the story of how her visit to an autism school had been a deeply "visceral" experience that "hit [her] in the stomach," she then began talking about one of the people that she had encountered there.

> So there was a boy in the class I was helping with, who didn't speak at all—very beautiful-looking (it's a cliché, but it's true—a lot of children with autism are); always had his hood up against sound; easily distressed; and spent a lot of time drawing in the air with his finger. And, just occasion-ally, if you sang to him, he would finish off what you were singing—with words, with sung words. But [he] never ever spoke any words (of speech or communication). Just think: if you're interested in this, I mean, how fasci-nating. How completely fascinating. And how completely tragic. I remem-ber, from my undergraduate experience, walking back out of the door and into [the] high street. And on one occasion seeing some mum, not exactly slapping their kid, but pulling their kid along and giving them a hard time about whatever it was—whining about something or other. And thinking that . . . it was so extraordinary that that ordinary child could communi-cate—you know, that this two- or three-year-old could whine to their mum about this, compared to these kids inside this school who, if they'd been able to do anything like whining, we'd all have been cheering and clapping. So, I don't know, it's hard to describe. Of course it's a very long time ago now. But I remember it clearly. I remember it very clearly. And it was partly probably intensified by the fact that I was traveling to these places on my own, and so I would go into a center for a couple of days, and go to a different place for a couple of days. The staff and children were very helpful and very likeable. It was still a very kind of, um, a very upsetting experience.

Looking again at these two extracts, it is striking how central bodies and feelings are to the two narratives. The entirety of the first story is carried along by the squeals and cries of one child, by the rocking back and forth of another, by the light filtered through the fingers of a third, and, finally, by the indefinable, visceral, upsetting feeling in the stomach of the researcher herself. It is also noteworthy that the interviewee interprets my question about emotions as a question about the body—and she affirms the specifi-cally affective nature of her experience on that level: "it's very visceral, yeah." In the second story, again her account of becoming an autism researcher is a story about bodies and feelings, her memory of the experience emerges from a narrative of singing voices, splayed fingers, slapped hands, beautiful faces, and the researcher's own feelings of confusion, fascination, and upset.

I draw particular attention to the triumvirate that appears at the end of her first story—where autism is something that the researcher is "passionate about," that she is "interested in," but that then finally "hits" her in "the stomach." There is very much a sense, here, and I encountered this in other interviews, of these three working in relay: that, once this researcher goes out into the tangible world of special schools and disordered children, her intellectual interest in autism enters into exchange with, and becomes thinkable through, affects, emotions, and bodies. Wilson (2004: 44) has argued elsewhere that the gut, specifically, is the boundary that mediates between an individual subject and the world around her. She quotes Michael Gershon to the effect that "the open tube that begins at the mouth ends at the anus . . . the gut is a tunnel that permits the exterior to run right through us" (ibid.). I cannot help but think about this story as a memory of the solid, exterior world of autism, with all its complexities, fascinations, and possibilities, beginning to run through the gut of the young researcher. Her early intellectual encounter with autism, as she recalls it, is not a simple placement of some diagnostic or biomedical object before the thinking subject. It is, instead, a more much complicated form of embodied understanding and exchange, where the researcher's intellectual interest in autism begins to almost literally pass through the stomach. This was a relation, I think—the clarity and force of her memory sustains this impression—*of* which she was quite aware and *to* which she was remarkably receptive.

Equally, in the second account, discussion of the science cannot go unaccompanied by an acknowledgment of a very visceral sadness and vice versa. "How completely fascinating," she says, on the one hand, and "how completely tragic," on the other. In fact, immediately following this story, which ended with the telling phrase "a very upsetting experience," she said, fairly quickly:

> Upsetting and fascinating in equal measure. And I suppose when I decided to do the PhD, I decided the fascination outweighed the distress. It wasn't motivated by some hifalutin idea that I was going to help.

I am not especially interested here—and I don't think there's a meaningful answer to it anyway—in whether her intellectual fascination or her feeling of upset were the proper instigators of the work that she went on to do. What I am trying to hold together is the way in which the mingling of these two experiences, of distress and fascination, of upset and interest—how their circling around one another and their doing so in the midst of such a heavily embodied account—how this begins to open up what is at stake in the conduct of an autism neuroscience—namely, the use of the body and of its emotions to sustain, generate, and animate an intellectual concern with the brain basis of these children's idiosyncrasies. In the service of this argument, I draw particular attention to the complex way in which between the first story and the second this researcher's intellectual concerns have acquired some identifiably affective commitments.

I tentatively take issue with her own conclusion that one of these ultimately outweighed the other—that the PhD only became possible to the extent that the upset was spent or that "hifalutin" ideas of empathy had been disbursed to clinicians and paraprofessionals. Even the essentially unbidden relation of this story bespeaks, I think, a much more complicated, much more longitudinal, relationship between this researcher's capacity to *think* autism and her willingness to *feel* it. I argue, ultimately, that it is her ability to trace her science through the two experiences, to memorialize them and articulate them together, that enables her to continue to push through the very complex work of autism research.

SCARS INTO DATA

But why would you trace a neurobiological account of autism through emotional commitments especially? To put it another way, can we infer any concrete relationship between (1) a scientific practice in which affective commitments are so manifest, and (2) the object to which that practice ultimately addresses itself—namely, in this case, a neurobiological account of autism? Because it seems to me that the recollections and feelings that I have recounted have some specific end in mind. I already quoted one scientist saying, "Here was this thing that I was really passionate about," and another saying, "If I was doing more of a pure science that wasn't . . . didn't have an end point that was kind of applicable—I'd sort of find that quite difficult." Next, we'll find another person who, in the midst of a blatantly emotional account, says, almost as an aside, "I want to know . . . what it is, fundamentally." But what *it*? What *end point*? What *thing*?

In fact, there is already a well-known discussion of the role played by emotion in the coming together of things, one that is also concerned with the troubled boundary between entangling and cutting, which is the work of A. N. Whitehead (1935, 1964, 1979). Whitehead's complex and somewhat confusingly rendered system is difficult to set out in such a confined space as this. But very basically, Whitehead's (1979: 19) metaphysics is centered on what he calls *prehensions*—moments of substantive attraction or connection between one entity and another. To put it at its simplest: for Whitehead, these substantive connections, or *prehensions*, between entities are basically constitutive of all objects and of all elements of all objects in this universe. But when he talks about the connections between *entities*, he includes, for example, a person regarding a chair but also, within the timber frame of that chair, the regard that the various molecular components of the wood have for one another.

A *prehension* is thus "any grasping or sensing of one entity by another or response of one entity to another: whether this takes the form of a stone falling to earth, or my looking at an object in front of me" (Shaviro 2009: 3). Whitehead (1979: 49) describes this *grasping* as a *feeling*, and it is particularly *feeling* that describes "the basic generic operation of passing from the objectivity of the data to the subjectivity of the actual entity in question." The crux for us is this: the moment of substantive and constitutive connection— which, for Whitehead, in its ongoing, processual character, is basically constitutive of all things—is a specifically *emotional* and affective moment, and it is always an instance of *feeling*. Whitehead (1935: 227, my emphasis) describes the quality of these interactions in terms of "the *affective* tone determining the effectiveness of that prehension in that occasion of experience." And thus any "occasion," for Whitehead (i.e., a successful instance of this movement, in which some concrete thing is achieved), "enjoys its decisive moment of absolute self-attainment as emotional unity . . . *the creativity of the world is the throbbing emotion of the past hurling itself into a new transcendental fact*" (ibid., my emphasis). If experience is the preeminent ontological fact, then "the basis of experience is emotional" (ibid.: 226).

What Whitehead is trying to breach, with this claim, is the fundamental division between subjects and objects that has bedeviled our capacity to conceive of, and talk meaningfully about, the concrescence of things for some centuries: this is his famous objection to the "bifurcation of nature" in which, according to Whitehead, a rigid separation was enacted between (in his example) the redness or the warmth of the sun and its chemical or molecular structure. Instead, Whitehead wants to say that it is not the case that one of these categories is proper to the sun, and one proper to the subjects, minds or bodies experiencing it—but that, in fact, what the sun *is* is the series of positive and negative prehensions between things that have the positions of subjects and objects at any given moment (and here, in positioning subjects and objects in temporary relationships to one another, he draws no distinction between electrons, particle waves, neurons, people, warmth, and so on) (Whitehead 1964: 30; 1979: 41).

More to the point, for my purposes, these relays are *emotional* in character; they have the quality of feeling. These acts of perception and prehension, Whitehead (1979: 116) argues, can "be conceived as the transference of throbs of emotional energy." What we begin to see with Whitehead, then— at least the sliver of his philosophy that I lean on here—is a disruption of the boundary between the feeling of something and that thing's constitution. As philosopher Isabelle Stengers (2011: 294) has pointed out, feeling (in the Whiteheadian scheme) begins to have the shape of a vector: "the point," she writes, of acceding to feeling, "is thus to take literally the common-sense statement 'this thing is present in my experience insofar as it is elsewhere,' and to construct its concept." Emotional, affective experiences begin to have the power of knowledge—real, concrete knowledge about what a thing actually is. To think with Whitehead, as Stengers (ibid.: 310) puts it, is to begin "transforming scars into data."

Here is where we begin to approach a more precise account of the relationship between these deeply emotional stories of autism research and the desire to nonetheless understand and describe the (to use Whitehead's language, *actual*) neurobiological basis of autism. In the Whiteheadian scheme, insofar as an actual neurobiology of autism might come into existence, it is as a product of the concrete emotional apprehensions of one another by its various elements—neurons, electrons, gluons, the white and gray matter of the brain, cognitive theories, scientific bureaucracies, learned societies, research participants and, not least, individual scientists themselves. In this sense, it becomes very understandable (to me at least) that scientists working at the coalface of neurobiological autism research, straining to understand this relationship, struggling to bring it into view, would begin to talk about their practice in such a deeply affective register.

I would go further: in talking through their attempt to apprehend the delicate connection between neuroscience and autism, and in reflecting on the fragility, still, of that understanding, would we not actually *expect* them to say things like "It was heart-breaking"; or "I found it completely captivating, and . . . terribly, terribly upsetting"; or "It's not just a job: you're dealing with people's lives"; or "I *love* the kids and I love the families"; or "It made me very excited"; or "I want to do this, I want to do this"? Would we not predict, of an autism scientist, a heavy investment in feeling and the presence of a notably emotional discourse? Would we not then anticipate, in her talk, precisely the kinds of descriptions of neuroscientific research that I have loosely gathered together here?

I am working hard to avoid overdoing this, for a couple of reasons. First, there is no comparator here, and I make no general claim about the specificity of emotional discourse to autism neuroscientists (patently, it is not so specific). Nor do I argue that good autism neuroscience has to be, or is always, a deeply emotional experience (again, clearly, this is not the case). But I *am* marking a reluctance, all the same, to see coincidence between the delicate and complex process of a *tracing* neuroscience, its awkward relationship to the gap between its theories on neurobiology and neurobiology itself, and the strikingly emotional and affective way in which many of these scientists talked about their own work. A second reason to be cautious—and here is one instance where we potentially run into the limits of the interviewing technique: I am making an ontological inference about scientists "working at the coalface," but I am doing so drawing only on a set of interviews. What role the affective situations that I discuss here *actually* play in the day-to-day conduct of autism science is impossible for me to say.

In the same vein, it is notable that the strongest invocations of affect come not so much from people recounting their day-to-day lives, but from broader discussions of what this science is like and how these researchers' biographies have gotten entangled in it.[3] It is important to note that, while I use phrases like "at the coalface" here and there, I am not making a claim about the affective status of day-to-day practice. In fact, I am not especially interested in the affective load of a particular day or a specific experiment.

What concerns me is a much wider affective economy, stretching across much longer expanses of time and place, and involving traffic between feelings, objects, people, memories, and experiments. I am interested in the relationship of that economy to moments in which *things happen*.

CIRCUITS OF LOVE

To concretize this, I move onto my final example, where both feelings about, and the constitution of, the brain basis of autism are very much in question. It loops back to where we began this account: with feelings of love.

One young researcher told me:

> It was from a theoretical stance, really, that I was interested in autism. But then I started working with two children with autism—one who was four, and one who was five. And both had very little speech. And I did applied behavioral analysis [ABA] with those kids. I did therapy once a week for, like, I think four or five years [. . .] so I was quite involved with the families and the kids, and their schooling. And I saw huge improvements, in kids who didn't speak at four to five, and didn't really communicate very well at all, to then being eight or nine and [who then] could converse—not in a grammatically correct way necessarily, but [they] could make themselves understood. And also there were stark differences in these kids, which is quite, I think, characteristic of autism in general—just the variability, in that one child in particular, was so . . . he was just . . . he just wanted . . . [*here she struggled audibly*] he wanted to sp— [*very quickly*] he wanted to have friends. He wanted to interact with other kids. He just didn't know how to.

Already, here, is the now-familiar juxtaposition between a prosaically scientific account and something that is altogether more self-consciously humane and emotionally committed. In the space of just a few sentences, we are discussing her "theoretical stance" on autism and the way that some clinical differences are likely to be "characteristic of autism in general," but we are also discussing an unambiguously poignant and even upsetting image. This is the position of the child she was working with, who had gotten to a stage where he really wanted friends but still didn't really how to get or keep them. This researcher does not name her upset as some of the previous interviewees did, but I strongly remember this interview being a bit uncomfortable and, to be honest, unexpectedly so. I seemed to be asking more awkward questions than I had planned.

As I have tried to render in the extract, there are a lot of gaps and silences on the recording when I listen back to it now, especially in places where you probably *would* name the emotion in question. There are even a couple of places where this very clever, accomplished, and articulate scientist is almost talking quietly to herself. "Was [this child] kind of articulating a desire for friendship?" I asked.

> Yeah, yeah, oh, yeah-yeah-yeah. Whereas the other child wasn't particularly interested in other kids. And his language didn't take, interestingly, as much as the other child's did. Because he just didn't interact with those other kids. They were both in mainstream schools. Um . . . but anyway that really got . . . I wanted to know why . . . I wanted to know why there was such variability in the autism spectrum, um, in terms of their outcomes. And . . . what we could do about it, I guess; what kind of factors might actually determine their outcomes. So might it be intrinsic child-related factors, like level of language to start off with, or IQ, or personality, temperament, or . . . general personality—or whether there were more kind of environmental or extrinsic factors, like how much intervention they've had, or the type of schooling they've got, or the type of family structure they've got. So I kind of got interested in that, and what we could do about it. I guess what we can do to try and . . . ensure that children with autism reach their potential, basically.

"Yeah," I said, "because that's a bit heartbreaking—the kid who wants to have friends but doesn't know how." She went on:

> Oh, I know. I mean that's what it was—I mean, the school he was in was lovely. So, that's just luck in some respects—he could not have been in a school that was lovely. And of course this was in primary school. When you get into secondary school, it's much more . . . it's hard for everybody, not necessarily just people with autism, but everyone.

She then steered away again from these kinds of topics (schools being lovely; kids having a hard time) and spoke in some detail, and again in that much more obviously *scientific* vein, about differences in autism. "Those are such different things to articulate," I said, drawing attention to this. "So that early interest in 'theory of mind,' as a kind of theoretical idea to explore. And, then, from that to this almost quite visceral kind of ABA [applied behavior analysis], very close interaction with these kids and these very emotional kind of difficult things. I mean—I guess I'm wondering, how did you transition from one to the other, or how did you negotiate the tension between those two things?" She replied:

> In some respects, there wasn't necessarily a tension. My research was on theory of mind—but also on other cognitive skills of kids with autism. And what I realized from my observations of working with the kids, but also during my research and the results of my research, was that these kids did have problems in theory of mind, which limits the sorts of interactions they can have and understand. But they also have additional weaknesses and strengths as well. And that also places them with advantages but also disadvantages as well. And so that was the tension, I think—[it's] how are we going to explain how these children can negotiate their social interactions, but not just their social interactions, but kind of their everyday lives . . . um, and, theory of mind wasn't up to explaining those difficulties.

From memory, I heard this as an unwillingness to acknowledge the obvious polarities in her account. In fact, in one sense her answer is quite unrelated to the question but nonetheless takes the discussion onto fairly safe ground. Yet it was still apparent to me that the autism she was working with was involved in a much more complex system of apprehension and exchange— one that lacked a clear or obvious bifurcation between competing cognitive theories, neurodevelopmental diagnoses, anxious children, and curious, empathic researchers. At the time, I remember being frustrated at her refusal to confront this.

What I did not hear then, however, and which definitely strikes me now, is her focus on and insistent return to the cognitive and neurological specificity of autism itself—*not* as an evasion but in fact as an *elaboration* of this relationship. As she says, and I think quite correctly: relations between different ways of constructing autism, *"that* was the tension." She is very clear—although I am deaf to it in the exchange—that she is not interested in emoting for its own sake, or to convince anyone of her empathic and humanitarian credentials, but to "know why." In other words, she is trying to talk about autism specifically, but not as a way to get around emotion or to ignore it. In fact—and here is where Whitehead is such a good guide—being able to talk about autism in these firm registers might be the whole purpose of her openness to feeling in the first place. In the interview, thinking we had gone off topic, I rephrased the question in a (possibly) more palatable way for a biomedical researcher and asked if a therapeutic desire had always driven her work.

> Yeah, I think so, I'm an educational psychologist by background, so I guess . . . yeah . . . [*long pause*]. I guess I want . . . [*another longish pause*], I'm quite, um . . . [*almost talking to herself*] it's not empathy . . . I guess empathy, with . . . I really get on with my families, so I've just written 120 Christmas cards to my . . . all my kids [*she laughs on "kids"*], all my autistic kids in [my home city] and here—just, you know, I do this every year. So, I get on really well . . . I love . . . [*she catches herself here, but then repeats quite definitively*] I *love* the kids and I love the families, and I think I feel that kind of rapport with them, so there's the sense that I kind of want them to do . . . I want them to do as best as they possibly can—because I can see that they have potential, and so, I think that's what drives a lot of my work.

As I noted earlier, the emergence of love is really striking here. I remember being a bit taken aback at the time—and I think she was too, given how much of an audible effort it was to articulate its presence. In some quite pro-saic ways, feelings of love should not be surprising. The people I interviewed, many of them (like this one) relatively young women, many at relatively early stages in their scientific careers, could not afford to be at all innocent of the emotional and pragmatic complexities of familial love (Donald 2012).

There is more to be said about the specific role of love here, however. Just as it was in the previous account, the move here is very obviously circulatory, as the narrative moves from theoretical stances, to love, and back out again. Just like the hurried claims to something beyond mere empathy described earlier, this psychologist's next words were:

> But at the same time, I want to know why—what it is, fundamentally, about kids with autism that is different to typical kids, so how do they perceive the world, and view the world that might be different to us . . . and how . . . and what we might do to ameliorate any differences.

I have said that accounts of autism neuroscience are often shot through with thick reports of sadness, upset, love, anxiety, and even pride. I have said that scientists who are methodologically open to affect are not just emotional for its own sake; in fact, their capacity for feeling, and for working through (and with) emotion, is heavily implicated in their commitment to working across an unbifurcated nature. This move is remarkably explicit here—we see autism research as heavily invested in the simultaneous passing by of thinking and feeling: that passionate attachments ("I love the kids") are not separated from intellectual interests ("that's what drives a lot of my work"). But also, again, this researcher's basic concern, as she is very keen to stress, is the *why*—what, exactly, is this thing underlying all these problems?

I am now disinclined to think about this as I did at the time, which was to assume that this scientist focused on the *why* in order not to talk too much about an embarrassing surfeit of affect—that is, that she "love[s] the kids . . . and the families." My feeling now is that the *why* of autism is actually what this whole system of affective labor is directed at in the first place. I suspect, in other words, that the specific, dry, and technical issues about the objective make-up of autism that skate endlessly across the top of these accounts are *not* simply a way to avoid talking about love; they are there, in fact, precisely to explain it.

DIFFERENCES

I end on this interview for a couple of reasons. Most obvious is the potency and frankness of the account. If I started this chapter with an embodied refusal of scientific distance (surprising enough to me), I end it with an even stronger affirmation of the affective load of autism neuroscience—with even more freighted stories of empathy and love. I have also ended it here specifically because this affirmation takes place in an *affirmative* register. There is certainly talk of problems and of heartbreak here (the latter introduced by me, in fairness), but we are no longer moved only by drawbacks, problems, and the concerns of parents (as real as these are, for many). Here, in addition, is a sense of exuberance in and love for "the [autistic] kids" that this researcher worked with, in their own rights. I have tried to stick closely, in this chapter, to what this data might tell us ontologically about how autism neuroscience happens. I am reminded nonetheless that another distinction is at play in these affective moments, which I gestured at briefly earlier in the chapter. This is a distinction between, on the one hand, feelings of distress at what someone has lost and, on the other, expressions of happiness at what someone is. There are sharp political questions embedded in this distinction, which need to be acknowledged, if not comprehensively addressed.

This is difficult territory. Not least, there is a risk of overinterpreting relatively small amounts of interview data. I am already at the edges of what I am willing to draw from the appearances of an affective tone in my conversations, and I am reluctant to get into a wider conversation about the politics of disability on foot of them. More to the point, we don't need to be committed affect theorists to wonder at a sharp split between the negative and the affirmative; not least, the final account, which I have here marked as affirmative, retains a sense of disadvantages and problems that can impinge on an autistic life—on things that may yet be in need of amelioration. By the same token, a story like the one I began with, which recounts a memory of a very "upsetting" experience, still goes on to talk about beauty and fascination, about moments when an otherwise silent child would begin to sing, unbidden. How are we to cut the negative from the affirmative in these memories? What would we gain in the attempt?

It is also difficult because I am unwilling to sit in judgment on these accounts. I wholly understand (to the extent that I *can* understand) why it is that many people who identify as autistic or as being on the autism spectrum are weary of being pinioned by rhetorics of suffering, of being understood via problems that need to be fixed, of being figured as the cause of parental and wider familial distress. But a scientist's feeling of upset in her stomach is no less real for that recognition; and it is no less socially, politically, and ontologically significant for all the complexity that it comes freighted with. Such feelings indeed are not necessarily extrinsic to forms of political mobilization in autism. As Martine Lappé (2014: 323) noted in her work on the ambivalent operations of care in autism research: "temporal, affective and material dimension" of research can generate "new forms of community and sociality within and through science that deserve ongoing attention." The bioscientific mode, for all its dour deficit-accounting, might yet have something hopeful to contribute to a politics of difference.

At stake here, of course, is a topic that I have gestured at but largely shied away from in this book: the topic of neurodiversity. As Pier Jaarsma and Stellan Welin (2012) have pointed out, there are generally two arguments embedded in claims to understand autism from the point of view of neurodiversity. One is that autism marks a real, biological difference from "neurotypical" states. In this sense, the claim to neurodiversity is quite strongly opposed to social-constructionist claims about autism, even to the point of sometimes being more committed to biological differences than the actual biologists (see Ortega and Choudhury 2011 for a discussion of the "neurorealism" of neurodiversity). The other is that the difference in cognitive style produced by this neurological difference is not a problem to be solved or a deficit to be cured—it is rather a variation to be understood, accommodated, and respected, in its own terms. As the autistic advocate and writer Jim Sinclair (2013: 2) put it: "Autism is hard-wired into the ways my brain works. . . . I know that autism is not a terrible thing, and that it does not make me any less a person."

In these debates, we often see discussions around neurodiversity placing—however misleadingly—autistic self-advocates on one side of an argument, and clinicians, scientists, and parents/family members on the other (see Hart 2014 for an important nuancing of this setup). It becomes immediately clear, within this context, how the affective accounts I relate in this chapter, which I have read in a broadly sympathetic vein, begin to look troublesome. These are accounts, after all, rooted in senses of distress, upset, and sickness in the stomach, offered up by (I assume; I never asked) broadly neurotypical people. However much the stories I have related here are at odds with a stereotypical view of cold scientific distance, we are still often caught up in a traditional narrative of deficit, sometimes unable to see how autism, or any other atypical neurodevelopmental experience, might unravel as other than misfortune.

The problem I have is that the distinctions and binaries in which these important debates are couched (the medical and the social; the neurological and the political) are the distinctions that I am trying to hold in abeyance.

To wonder at my interviewees' commitments to a medical model of autism, and/or to their (typically) neurotypical discourse of loss, seems to run counter to the spirit of what I am doing. That decision has political consequences of its own of course. Not least it seems to set this book against a preference for the social model of disability (broadly construed) in related literatures (see Shakespeare and Watson 2001). Indeed, a reparative attitude to the neuroscience of autism raises the question of whether there is not potentially some gain in suspending this valorization of the *social* over the *medical* as such and even in troubling a preference for a language of accommodation over a language of deficit. Any possible political alignment between autistic people, their allies and families, clinicians, scientists, and third-party organizations, across such a question, is intensely complex.

I have raised but left quite unanswered important questions about how we are going to think (and who "we" are to think) about the relations of diversity and deficit in the absence of a line between the medical and the social—how in the future we are going to plot the interrelations between a science bent on intervention and a politics committed to acceptance. Put another way, there are important questions, still to be addressed, about how we are going to think the politics of disability through relations of reparation to high-tech bioscience, and to clinical practices, when that politics has consistently gained its momentum by sitting athwart those sciences and those practices.[4] I will not address this question directly in what follows, but I retain it as an important cartographic device, as I start to map the relationship between the arguments in this book and the new brain sciences more generally.

4

COMPROMISING POSITIONS

Psychology and Autism between Science and the Social

IN THE PROLOGUE TO HIS BOOK *MAKING UP THE MIND* (2007), THE WELL-known British neuroscientist Chris Frith discusses his undergraduate decision to switch from studying physics to an entirely new offering: psychology. Frith (2007: 1) describes, with seeming regret, how the switch immediately placed him lower down the scientific pecking order: "I have continued to study psychology ever since," he assures the reader, "but I have never forgotten about my place in the hierarchy." As the book progresses, however, it becomes apparent that, if sincerely intended, such modesty might not represent the situation in its entirety. Indeed, precisely illustrating why this hierarchy is no longer so potent seems to be among the major purposes of Frith's reminiscence. "Much has changed in psychology over the last 30 years," he points out: "We have borrowed many skills and concepts from other disciplines. We study the brain as well as behaviour. We use computers extensively to analyse our data and to provide metaphors for how the mind works. My university badge doesn't say 'psychologist' but 'cognitive neuroscientist'" (ibid.: 2). The section of the book in which this recollection appears is called "Real Scientists Don't Study the Mind." As the argument develops, it becomes apparent—perhaps inevitably—that Frith's intention with the title is either ironic or historical. His core claim is precisely the opposite: "big science" he says, in a passage pasted underneath a picture of an MRI scanner, has "com[e] to the aid of soft psychology . . . we no longer need to worry about these soft, subjective accounts of mental life. We can make hard, objective measurements of brain activity instead" (ibid.: 12, 15).

CHOREOGRAPHY

What are we to make of such deft autobiographical and intellectual chore-ography? How should we read this compelling witness to the changes that have taken place in British psychology and neuroscience in the last decades of the twenty-first century? What's going on here, of course, is in part a story about the different histories and trajectories through which people, concepts, and approaches have come to be understood as *neuroscientific*. As I noted briefly in the introduction—and in this chapter it really matters—the majority of people I spoke to were neuroscientists in a fairly unproblematic way; but most of them were also *cognitive neuroscientists*, which usually meant they had an affiliation to, or training in, the discipline of psychology. It's simplistic, but as a heuristic, and in purely disciplinary terms, we can imag-ine *neuroscience* as the centerpoint of a host of overlapping circles: within this agglomeration there are circles for *psychology, psychiatry, neuroanatomy, molecular biology, biochemistry, computer modeling,* and so on.[1] Somewhere in the general area where the history of psychology runs into the study of the brain, we could roughly pinpoint the discipline that a lot of my interviewees subscribed to, which in 2007 was marked on Chris Frith's badge: "cogni-tive neuroscience."[2] So far so good. But to this one-dimensional account, we would quickly have to add another dimension: time. This would allow us to consider the idea (apparently held by some) that the neurobiology of mental states is in fact best modeled virtually, so computer science starts to move around and into this cluster of things we're calling cognitive neuroscience, subtly altering it as it goes.

There are two important points underpinning this slightly stupid descrip-tion: (1) the discipline that many of my interviewees subscribed to—cognitive neuroscience—as an intersection, more or less, of the study of psychological faculties and the study of the brain has a strong element of temporal mobil-ity and contingency to it; (2) skating across that temporal movement, and in some ways driving it—here especially is why I draw on Frith—is a very specific claim to the unambiguous status of "science" from people who were once based in psychology, but who are now starting to inhabit this intersec-tion instead. I said already that one of my broad interests is in enacting a new kind of conversation between the social sciences and the neurosciences. I have pursued that by bringing to light some of the nuances and ambiguities that lie at the heart of neurobiological research. I have so far told this story through an account of neuroscientists delicately (but doggedly) tracing the neurobiology of autism through various scenes of definitional, epistemic, and affective complexity. I have emphasized the novelty, the dexterity, and the care with which autism neuroscientists are actually able to recognize and work through such complexity. I have shown them drawing very different kinds of knowledge and commitment together; I have described them as working with contradiction, and as refusing to be cowed by ambiguity. I have tried to demonstrate how they are nonetheless always, still, carefully working toward a coherent, convincing neurobiological account of autism.

In this chapter, however, that story runs up against something of a limit. Because amid all this complexity, sensitivity, skill, and care, there was *also* very often in my interviews a rather jarring (and very explicit) characterization of psychology, under neuroscience, as (now) an unambiguously hard and objective *science*. Such invocations were neither incidental nor ritualistic. Indeed, they were quite specifically scient*istic*, in that the image of a neuropsychology was generally the basis of a knowledge-politics—one that made sharp distinctions between, on the one hand, the classical virtues of hard science and, on the other, the inadequacies of the fluffily subjective. This commitment was underwritten by a collective representation of psychology as having only recently emerged from decades of mystification but having become now, thanks to a series of technological and conceptual developments, the expression of an uncomplicated materialism, addressed to a measurable organ. This representation took it for granted that within this organ lay a whole series of valid and specific psychopathologies, which would, sooner or later, be robustly characterized by quantifiable biomarkers, thus becoming subject to intervention, change, and so on. I am caricaturing a bit here. Yet over and over again, I was given to understand the work of locating autism in the brain as the practice *also* of a hardening science— one that would eventually banish the soft and subjective from the space of psychological knowledge-making.

This was odd. I have perhaps rather belabored my claim that within a broadly conceived neuroscientific discourse, autism tends to resist simplicity: within the accounts of autism researchers themselves, and (mostly) without shame, autism is depicted as endlessly entangled in not only medicine, psychology, and the brain, but also, as the medical humanities scholar Stuart Murray (2011: xiii) has described it, in a much "wide[r] fabric of narrative, representation and characterization." If I learned anything from my interviewees, it was that nothing about the neuroscience of autism was straightforward or easy. How, then, are we to understand the enrollment of this practice into a simplistic caricature of a *scientific* neuropsychology? How does such an enrollment relate to the larger picture of neuroscientific ambiguity and care that I have been trying to erect here? Where do these claims sit within a longer history of vacillation between psychology and science? What is the significance of the neuroscience of *autism* in the mediation of that relationship?

My strategy is to think about all of these questions simultaneously from a single direction—namely the temporality of these researchers' talk about psychology as a science. When such invocations of science came up, they were often introduced (as Frith introduced them) as moments in the interviewee's own career or lifetime, or via reference to the discipline's collective memory of its own shameful, mystificatory past. I focus on temporality because it allows me to draw in an interpretive frame from the philosopher Georges Canguilhem, via Dominique Lecourt (1975) and Nikolas Rose (1996a)—namely, that histories of psychology have a tendency toward recurrence; that they are often ways of arranging the past into justifications of the contemporary; that they are formed, and reformed, to demarcate and regulate the boundaries of psychology's present as well as its future.

If this rush to claim the status of *science* might run counter to the careful work described already, I nonetheless want to work the data in this chapter through a different but related portrait of a tracing neuroscience, one not yet discussed, and this is the relationship of the trace to a particular kind of epistemic boundary-making (Gieryn 1983). To "trace autism up," as a scientist, is to think across borders—borders like the one between intellect and the body or between biological truth and diagnostic convenience. But a tracing autism also sometimes runs the researcher, quite directly, into the borders of some very different epistemic cultures (Knorr Cetina 1999)—and this is not always easy. Is it possible that when an interviewee says that she was relieved to learn, as an undergraduate, that psychology was a *proper* and *scientific* subject, that it was not about "faff"—that this is not a holiday from complexity but is actually a way of articulating, working through, and dealing with the intricate web that is now in the process of being formed from some very different epistemological positions?

One boundary that I have in mind is that between a quantitative science of psychopathology, on the one hand, and the ongoing incursion into this milieu of an unruly and unfathomable *social* world, on the other. It is the social that bubbles along under this scientism, for two reasons. First, although the long-running story of psychology-as-a-science reaches something of a climax with the institutionalization of cognitive neuroscience at the start of the twenty-first century, there has nonetheless come, at the beginning of that century, something of an unexpected coda. The brain-imaging era has indeed produced a renewed scientific confidence (Kandel 2007). Yet the great flowering of biomarkers, diagnoses, and drugs—which were to quickly come from access to a realm "beneath" ordinary mentation—has never really come to pass (Hyman 2008). This disappointment is strongly marked by the reemergence within psychology of an etiologically and diagnostically salient *social* world, an influence remarkably difficult to account for in a laboratory science, and which wraps itself around neuropsychological accounts of distress via the familial and cultural salience of epigenetic effects, gene-environment interactions, and so on. Second, I said in chapter 1 that autism is a disorder (diagnostically) predicated on particular ways of thinking about social interaction and communication, and also that autism has always been particularly hard to disaggregate from the complex movements of the societies that render it visible. One of the most striking things about autism as a research entity is that crossings of nature and culture, which are so often embroiled in psychological accounts, are very much on the surface. I want to explore the possibility that working the border of psychology and science might be an occupational hazard of tracing *autism* particularly—as neatly bounded studies of brain run headlong into the claims of culture and representation.

What I am pursuing here is psychology conducted at the sharp edges of science and the social. We realize, when we rub against those edges, that the troubled boundaries of a tracing neuroscience are not always negotiated with ease and grace; that sometimes such work is marked by quick and crude attempts to rearticulate, reerect, or even just find oneself within, some grounded marker of distinction. But if these questions and difficulties are especially evident in conversations about autism, I want to tentatively suggest that this description might be applied to the ongoing emergence of psychology as a *cognitive neuroscience* more generally. As we proceed, keep in mind the notion that the psychological future might not lie in a movement either toward or away from (to use Frith's term) "big science"—but rather in learning to trace, much more carefully, the connections between what matters in the scanner and what takes place in the world outside.

FREUD AND FAFF

"Psychology," said a young lecturer to me once, "when you [thought] of it at a school level, [was] sort of Freud and faff, and not really scientific, [whereas] what I wanted to do was, you know, do something scientific about the brain." This came a bit out of the blue: we had been talking about her PhD, and how it led to doing projects on autism—with no mention at all of science as such, still less of Freud or psychoanalysis. In fact, as these conversations went on, I realized that such enthusiastic invocations of psychology as a science, as well as dismissals of approaches deemed *not* scientific, often arose rather suddenly—arriving without a great deal of foreshadowing and usually without obvious reference to whatever else was going on during the interview. They tended to circle around a small set of common images—referring to a strong intellectual preference for physiology and the brain, explicitly differentiating the speaker from an older era of psychological flakiness (often represented, as here, by a caricature of Freud), referencing a desire for quantification, or sometimes even leaning on an image of human animality. She went on:

> I was really pleased when I discovered that, actually, you could do experimental psychology, that it was a proper subject that you could study and things. I guess I'd always been interested in these kinds of questions about social cognition, um, but it took a while for me to get round to finding a way that I thought those questions could be addressed scientifically, rather than in a sort of arm-waving kind of sense.

I am interested in how much of this lecturer's earlier nuance goes missing here (this was the person who told me that brain imaging "bears no relationship to reality" in chapter 2): ways of thinking about mental phenomena are split between "faff" and approaches that are "really scientific." Freud, meanwhile, whatever the pretensions of his early desire for a "scientific psychology," is dismissed out of hand as an "arm-waver."[3]

The nonobvious "arm-waving" is an interesting image. It carries the suspicion of rhetoric, an attempt to convince and to build argument, the location of an argument within the body of an individual, *personal* knowledge—all those images of an eighteenth-century proto-science discussed in chapter 3. What's also interesting is this interviewee's evident relief at the discovery of an experimental psychology that was, by contrast, a *proper* subject—by which she meant one that was *really scientific*. This is what I mean by the invocation of a scientistic imaginary in these conversations. The use of a word like "proper" is a clue: this is memory working to mediate what counts as legitimate practice in the present; it shows the art of recall as both self-conscious recurrence and present-centered program (Rose 1996a: 43).

Consider in the same vein the following discussion, which comes from a more senior scientist, trained in the era before brain imaging but also somehow haunted by Freud, the ghost of psychology's shameful, half-remembered past. Remembering the advice to switch from an undergraduate tutor, she said:

> I had actually not wanted to do [psychology], because the only psychology I was aware of was Freud and Jung and that, which I'd tried reading and found irritating. But then I realized that it was actually a much more sort of scientifically based [at her undergraduate institution]—it was experimental psychology, it was linking in with physiology that I found . . . I was quite interested in . . . I suppose I've always thought, "Are there physiological differences that explain why people are different?"

The recourse to a specifically physiological approach as the only one acceptable to her marks something important, as does the rhetorical deployment of physiology to mark the division between practices that are either (a) Freudian and irritating or (b) experimental and scientific. Later on in the conversation I asked her: "Why is it important that it would be a physiological psychology or one that's in touch with physiology?"

I don't know, um, the answer to that. It's just . . . I think it's just my, perhaps a somewhat reductionist bent I have, um . . .

"Right," I said, struggling to properly articulate my question: "Or what irritated you about Freud—let me put it that way."

Oh, I just found it all untestable and vague and um . . . I just felt there was no way of testing it. I mean, I was right from the start, I was very interested in the idea that if you've got a theory it should be testable. And I thought he, you could make up . . . at the time, this was when I was about eighteen, I'd just given up religion. And Freud struck me as part of the same sort of, you know, you believe what the great man says, but you don't sort of have any way of really testing it.

What marks this account is a specific embedding of this scientist's own narrative in references to the physiological body, confessions of a specific tendency toward reductionism, and, right at the end, a remaking of the same division she erected at the beginning of her story—here marked as a split between approaches that are either *testable* or *religious*. It is telling that she positions her own entry into psychology at the precise moment in her life in which she had "given up religion." The coincidence of the chronology is, again, neither here nor there. What I hear, specifically, is personal memory put to a very specific use: to embed the interviewee's own intellectual career within a concretely and unapologetically scientific tradition.

Here is another account of psychology, given in the same sort of physiologically prescriptive vein. In fact, this interviewee had started our interview, without any prompting from me at all (this was quite early in the process), with a quick and blunt claim about science, saying of his undergraduate degree in psychology:

It was very much a branch of the biological sciences and presented as that . . . . It used to be in the [humanities] faculty and then you did a degree in philosophy and psychology, and then for a few years they moved into science, which happened at the time I was there—where you did physics, chemistry, biology, mostly biology—and then moved to graduate specializing (it was then a four-year degree, which was terrific) and, uh,

my degree is really natural science. [. . .] Most of the students hated it. You
must remember, in those days, you know in my final-year cohort there
were fourteen students. These were the good old days of when university
was a kind of elitist experience [*laughs*]. And I think I was the only one
who said, "Yes, I think this is the way psychology degrees should be." And
I still think that.

What intrigues me about this contribution is the self-conscious effort not only
to present himself as a biological scientist but also his desire to let *me* know
that he approved of this designation, and that in fact he had approved of it
even at a time when such approval was neither popular nor profitable. Later,
he said:

I think we are embodied organisms and while I am not a complete biological reductionist by any means, you cannot talk meaningfully about the way
people act in their environment without understanding the fact that we are
biological organisms. It's an absolutely necessary part of understanding that.

We should note that there is an important qualifier here ("not . . . by any
means"), and I will be faithful to it when I reintroduce this interview later in
the chapter. Let me just note here that not only was he keen to stress that the
object of psychology was, specifically, a *biological organism*, but he went on
to add a scatological, and even animalistic, coda to this reasonable-seeming
description:

Without wishing to be gratuitously rude or whatever, but, you know, we
have to piss and shit and sleep and have sex and stuff like that. And there's
a point for any of us in which those kinds of things take over. And, uh,
our life is kind of about managing—Freud in a sense was right: we have to
manage the social niceties, and the fact that at some point I feel, you know,
during this day, I feel like I'm going to have to take myself off to the loo and
take a shit.

I am slightly struck upon rereading, here, by the presence of a more sympathetic reference to Freud than we have had previously. At the same time, and
in the midst of this relatively wide-ranging and sympathetic view, he insists

on a laceratingly biological image of human animality. This image is not only about shitting in general but quite deliberately (I think) breaks a social taboo by reminding me of his *own* biological need at some point during that day to go and take a shit. This is not to position animality as a necessarily narrow or reductive position (Wolfe 2003). But it is precisely via a reference to a sense of biological animality, and a memory of his own early reconciliation with it, that he justifies a deeply natural-science approach to psychology. It is just as he says: "Those kinds of things take over."

These are only some of the more vivid accounts. In fact, these kinds of memories and claims came up again and again, as did other memories, often secondhand, of a different era in psychology's formation as well as fears of unscientific approaches pursued in other places. Another junior researcher said of her first encounter with psychology:

> The psychology department [at my undergraduate university] is a department in the natural sciences, that also influenced [my] approach very much, and I also like the biological view of psychology and the natural sciences view of psychology very much that they have there.

Another, in response to my puzzlement at the persistence of this emphasis within my interviews (this was well into the interviewing process, when I was starting to assemble an image of ambiguous, uncertain neuroscience) explained to me, rather patiently, that psychology had become "more biologically focused, based in the brain. So a lot of psychology departments are now called cognitive neuroscience." Yet another told me:

> Um, amazingly I think, in this country, often, if you get a diagnosis [of autism], you will be referred to [an institute historically associated with psychoanalysis and psychotherapy] . . . which amazes me. And I've spoken to friends who've gone there and of their experience, and I'm absolutely horrified by the things that they're told [. . .] but again, that's sort of my scientific bias coming in I guess here.

Or consider this account from a clinician, who was equally keen to root himself within a scientific and biological tradition, but who did so against a mirror image of time rather than place:

> I'm not sure whether . . . because this would be true for other disorders,
> where you know twenty, thirty years ago there wasn't the sort of brain,
> neurosciences-type of research, and now there sort of is and that must relate
> to a whole bunch of clinical conditions [. . .]. I've always had a relatively
> medical model [and been] happy to embrace sort of biology.

And there were still other ways of talking about this, too, that I have not
really considered here. A postdoc, for example, qualified her own intellectual investment in psychological phenomena as a quantitative interest only:
"I like physics too," she said. "Everything that deals with hard data or numbers, experiment, things that you can measure, that appeals to me." Let me
close this description with the following account, which comes from one of
the small number of doctoral students I interviewed, and who in fact after
an undergraduate degree in psychology was working toward a PhD in neuroscience specifically. She talked to me about her experience of studying psychology in high school, which in her recollection was quite an ecumenical
experience. In the exam you had to consider a fictional patient's psychological
problem from the point of view of a clinician, then take whatever approach
you wanted—including the psychoanalytic. The now-PhD student said:

> So when I did this and I started to look at all these different approaches,
> the biological one always gave me answers that I just trusted a bit more. It
> always just fulfilled me a little bit more when I thought about what's going
> on in the body [. . .]. But bizarrely, I don't know how I got so . . . fixated on
> this idea that I needed to know about the brain. But for some reason I did,
> and so then I [*laughs*] I like pestered the people in the biology department
> [at her undergraduate university] to let me do some of the courses on, you
> know, receptors and neurons and stuff like that, and so they did. Although
> it wasn't actually on offer.

This memory might actually reflect a shift in the history of British psychology—one in which even psychoanalysis is still part of the discipline's visible
intellectual hinterland, although it seems somehow intrinsically unsatisfying.
But it is equally a moment in which a young would-be researcher is drawn
inexorably toward a specifically biological approach, for reasons that she can't

quite articulate (and this interviewee's loose grounding of her relationship to biology in a sense of trust, especially, is interesting in this context).

It is certainly remarkable that so recently her (prestigious) psychology degree program did not include a neuroscientific component or much of a brain-based focus ("it does now," she assured me). But my interest is in her clearly articulated sense that to *be* a psychologist was to seek out brain explanations all the same. Despite an institutional and pedagogical lag, the science of psychology had become—to this interviewee at least—second nature. "When I really can't understand something," she said to me later on, "then I often, like, try to take it down to: okay, what would the neurons be doing?" In what follows, I think through the ground on which this self-conception rests. Why, in this period, amid a cohort of autism neuroscientists so well versed in the strange entanglements of their own pursuit, would a nascent career be *so strongly* narrated through a commitment to these unabashedly organic and scientistic tropes? Why would it be located within such an explicit rejection of any major historical alternatives?

EVERYONE HAS HIS CLASSICS

In a 2010 article for *Scientific American,* and situating his analysis within a discussion of autism particularly, then NIMH director Thomas Insel lauded the emergence of approaches from the new brain sciences within the general study of psychological and psychiatric distress, which he took as an invitation to declare that now, finally in 2010, psychology and psychiatry had become fully and properly scientific.[4] If in the past, psychological and psychiatric analysis addressed itself to some purely mental and therefore speculative function, "today," wrote Insel (2010: 44), "scientific approaches based on modern biology, neuroscience and genomics are replacing nearly a century of purely psychological theories." Claiming a redefinition of psychiatric and psychological distress as a series of problems in "neural circuitry," Insel (ibid.: 51, my emphasis) argued: "From the scientific standpoint, it is difficult to find a precedent in medicine for what is beginning to happen in psychiatry. The intellectual basis of this field is shifting from one discipline, based on subjective 'mental' phenomena, to another, neuroscience.

Indeed, today's developing *science-based* understanding of mental illness very likely will revolutionize prevention and treatment and bring real and lasting relief to millions of people worldwide."

If we go with Insel (let's leave aside neuroscience's in-fact-happily-ongoing relationship to the subjective for a moment), it might be said that the accounts in the previous section simply reflect the march of history, that my interviewees associate themselves with an unproblematized science only because they are now, unproblematically, scientists. This is the explanation that would perhaps appeal most strongly to many of the interviewees themselves. And it has all the attractions of parsimony about it. But it doesn't work. As I have labored to make clear throughout the book, the neuroscience that I encountered during this project was almost always a complex, contested, *weird* intellectual activity—occasionally silly and arrogant, certainly, and sometimes crudely reductive. But more often than not, when I spoke to neuroscientists, I found a neuroscience that was not only happy to trace neurobiological accounts across all kinds of intellectual, affective, and definitional borderlands, but a neuroscience that seemed to be in fact, quite specifically, *mining* those margins. To simply say that we are in the middle of a shift from the subjective to the scientific actually tells us nothing at all about either psychology or neuroscience, as I have come to understand these practices.

As I already noted, whatever their burgeoning facticity within (some) popular and journalistic imaginations, the neurosciences are still quite self-consciously hybridized from other disciplines (Abi-Rached 2008; Abi-Rached and Rose 2010). While some of the elements of this hybrid (physics and computational science) carry a lot of scientific and clinical capital into the new endeavor, other parts (psychology and psychiatry) are less well-endowed. This leaves studies of psychopathology a bit vulnerable: while the enrollment of psychological concepts into neuroscience has removed some of the whiff that has tended to trail behind the psychological sciences like a dust cloud, recent discussions might also lead us to ask whether that confidence has not in fact been a temporary development (Diener 2010; Neurocritic 2012; Bor 2012).

More recently, Insel's thinking developed into what is now known as the Research Domain Criteria (RDoC)—an initiative by the NIMH to, among other things, shift psychiatric classification away from reported symptoms and toward the measurement of biomarkers (there is much more nuance to this than I can convey here; see Insel 2014 for a fuller discussion). As the psychological anthropologist Elizabeth Fine (2016) has pointed out, however, the future mapped out by RDoC seems to unconsciously reproduce some of the history of research on, of all things, *autism*—such as the link with identity (because we are no longer interested in "disease" as such, but rather in the biology of an individual brain) as well as a kind of agnosticism about whether measured traits (or symptoms) get experienced as negative. History, again, intrudes: "This seemingly radical shift in how we conceptualize neurological normality and abnormality," writes Fine (ibid.: 178), "is still embedded in older assumptions."

All of which makes it difficult to read Insel without remembering Canguilhem's acid remark, made more than half a century ago, that "if one terms classical psychology that psychology which one is proposing to refute, we must say that in psychology everyone has his classics" (Canguilhem 1980: 44). Canguilhem had his own ax to grind of course. But his observation draws our attention to a quality of psychological talk and psychological memory that I will lean on quite a bit in what follows. The suggestion, rooted in Canguilhem's historical epistemology, is that histories of psychology tend to work within a double move. In one sense, the history of psychology might be written as a straightforwardly teleological story of the establishment of particular investigatory practices, the emergence of a laboratory space and a set of techniques, the formalization of methods, the emergence of testing and statistics, the growth of professionalization, the emergence of the major paradigms of the twentieth century, and so on. This kind of account tends to begin with Aristotle, passes over Locke, runs straight to Wilhelm Wundt and to William James, skips lightly across the psychoanalytic detour—before landing more or less squarely on the quantitative biological science that we know and love today (I am caricaturing, but you get the idea).

In a second sense, though, such histories are set up and told quite specifically to arrange the past into a reasonable account of, and justification for, the speaker's claim on the present. The dominant trope of such recurrent histories is often the (eventual) settlement of psychology as a *science*. "The behavioural sciences have left the armchair and entered the laboratory," wrote the psychologist Michael Wertheimer (1987: 156) at the close of his *Brief Introduction to Psychology*. "Reliance on wise, experienced minds, equipped with oratory and quill pen and paper," he continued, "has given way to reliance on impersonal scientists with their precise measurements, their cold numbers, and their electronic computer."

Writing for a textbook in 2003, Alfred Fuchs and Katharine Milar (also psychologists and neuroscientists) suggested that psychology finally "appeared to be less self-consciously concerned with the status of psychology as a science and more concerned with the kind of science psychology was to be," which is both a neat acknowledgment of this long-standing anxiety and the role it has played in psychological historiography but also still, of course, an expression of it too (*now* we are scientists). In his work on the history of statistics, historian of science Theodore Porter (1996: 209–12) has drawn attention to the emergence of statistics in psychology in the early twentieth century, showing how, just as neuroscience and genomics would function a century later, "up-to-date statistics became a mark of self-consciously scientific experimental psychology . . . researchers were urged to follow statistical rules as a matter of scientific probity, and to feel guilt if, for example, they reformulated the hypothesis after the data came in."

As Porter shows, this resting of psychology's scientific respectability on the "rigor and certainty" provided by statistics had less to do with statistics as such (then comprehensively split between the approaches of foundational statisticians Karl Pearson and Ronald Fisher), and rather more to do with a desire to locate, on whatever basis, some strongly scientific basis for this growing discipline. This simultaneous reliance on, and black-boxing of, a calculatory logic in order to ground the scientific credibility of psychology is a significant precursor to the more organic and biological claims that I outlined earlier. I well recall my own naïve surprise, during an early foray into a clinical brain-imaging site, upon learning that an fMRI brain image was really only a way of illustrating a data-set—that behind each thick, rotating, colored brain image, lay a two-by-two array of individual numerical calculations and directions. Indeed, STS scholar Anne Beaulieu (2000: 12, 63) has shown how the process of "biologization" that many scholars have described is just such a process of "digitalization"—of kneading hard, numerical figures into soft, organic edges.

The emergence of behaviorism is another way to think this circular, past-repudiating scientism, memorably described by Canguilhem (1980: 47; cf. Mackenzie 1977) as "the principle of the biological psychology of behaviour" based on "the definition of man himself as a tool." Or take the randomized control trial of our own era, which, as the historian of science Trudy Dehue (2001: 296) has pointed out, expresses "the aspiration of ruling by technique rather than tradition, of replacing the individuality of both the governors and the governed by impersonality." A century before both of these developments, we could look to the emergence of a claimable science of psychology through attempts to elide the personal characteristics of experimental participants. Kurt Danziger (1994: 74) showed how, for example, in studies published in the *American Journal of Psychology*, the percentage of studies naming the subjects dropped from fifty-four to twenty-four in less than half a century.

As another historian, Mitchell Ash (1992: 193), has pointed out, these ongoing, recurrent claims to science *now* should tell us something important about psychology and about the way that it occupies a peculiar situation in "the status system of the sciences." Psychology is, after all, "a collection of quarrelling specialities and schools, pulled to and fro between methodological demands presumed to have been derived from the 'exact' physical and biological sciences and a subject matter extending uneasily into the social and human sciences" (ibid.). Is not the neuroscientific-organic claim within my data only the latest turn of this wheel? Are these claims not then propelled by a long-standing tradition of oscillation within psychology—as well as an insistence, at each recursive bend, on being shocked (*shocked!*) at the subjectivity and crudity of the moment just passed?

My point is not to disrupt or to critique the claims of a scientific psychology. Danziger (1994: 2) has suggested that to talk about "a field like scientific psychology" is to talk about "a domain of constructions" and that "the key to understanding its historical development" would therefore "seem to lie in those constructive activities that produced it." But it seems to me that a more useful way to think about the claims of a scientific—and *scientistic*—psychology might be to focus on what such claims are actually trying to *do*, positively, in the present. Nikolas Rose (1996a) has proposed that we not disaggregate the way that psychologists want to represent their history from what is possible or desirable within a psychological rubric at any given moment. In other words, that we might begin to trace important and functional relations between the kinds of psychological knowledge that hold sway, the scientific histories with which those knowledges seek affinity, and the sorts of things and people that those knowledges help to bring into existence. This is what Rose (1996a: 42–43), following Lecourt (1975) and Canguilhem, has called the recurrent nature of the history of psychology—an ever-spiraling move in which the past invariably helps to "demarcate that regime of truth which is contemporary for a discipline." The history of psychology is in this sense neither artifice nor lie; it is rather a methodology, one that enables its practitioners "to police the present, but also to shape the future" (ibid.).

VEERING BACK AGAIN

Let me return to one of the interviewees I quoted earlier in this chapter, the charismatic and youngish professor who was keen to emphasize the pissing/ shitting/eating animality of human experience and the necessity of understanding this experience biologically. Within the course of the same monologue, though, there is another vision of psychology in play, albeit briefly:

> Psychology is about the behavior of the individual, and the individual in relation [*here he catches himself a bit*]—so, I mean that covers everything that I do, but I don't say that everything can be reduced to [the] brain, even though I do brain stuff. But, uh, psychology is at very interesting crossroads between systemic, societal, emergent explanations of why somebody does things.

This is interesting because my interview with this young professor does not at all read like an interview with a person at the crossroads of anything. Yet here emerges a rather different vision of psychology—a discipline whose objects are not especially given to a biological approach but that also have "systemic," "societal," and "emergent" properties that need to be understood. Rereading my interview with him now, in fact, this professor seems a bit back-and-forth on this question. Later in our conversation, he used the analogy of a broken computer to differentiate between the ways that a sociologist and a psychologist would individually approach a problem. He said:

> I mean, the computer is a physical thing, which has particular properties which are determined by its structure. And an understanding of that structure in a sense determines all the kinds of things right up from, you

know, trying to understand the impact of the internet on society or something. The internet is partly the way it is because of the social structures that make it possible. But it's also partly the way it is because of the strengths and weaknesses and characteristics of what it is made of.

Here again there is a sense of something else at stake, something that requires us to take account not only of whatever it is something is made of but also "the social structures that make [that thing] possible." The tricky part—I suppose, the *crossroads* he mentioned—is including these social elements such that a psychological account would be no less biological and certainly no less scientific.

Perhaps this use of the word social was a concession for my benefit. I came across similar expressions quite a few times throughout these interviews. In one case, a psychiatrist told me that the sheer amount of discussion and debate around categories within the *DSM-5* made many of his colleagues still "fee[l] like it's in the realm of social psychiatry." A social neuroscientist, discussing some of the particular difficulties of her own special interest, said that "in social cognition, the more you try to control the [experimental] situation, the less you end up studying social cognition because social cognition has to be very, very flexible." A young psychologist spoke about her dual training in psychology and sociology, saying:

> I was very interested in the brain, and I guess its functioning in isolation from social groups. [But] actually it's even now quite difficult for me to think about there being a stark boundary between psychology and sociology. Because I think now a lot of what we know about human, about, um, psychological processing, is strongly influenced by being in a group or not being a group. And so, it's not very easy to draw a sharp boundary between them.

This relatively recent revelation ("I think now") of the difficulties of drawing a "sharp boundary" between the object of a *psychology* and the proper concern of a *sociology* is precisely what I am trying to get at. What is common to these contributions—whether for good (as in social cognition, a burgeoning field premised on it) or ill (in psychiatry, a medical specialism still grappling for organic respectability)—is a reemerging sense that, whatever the claims that *big science* has made on *soft psychology,* a lot of research on the neurobiological underpinnings of psychopathology is still not so easily disentangled from the complex machinations of the social world.

In all of these different claims to science, one vital thing holds them together: whether based on the ability to locate psychological phenomena in the laboratory, or the capacity to quantify these phenomena, or the development of tools to leverage them for averaging and prediction, the pull that science exerts on psychological classification is *away* from the social context in which the individual finds herself. As Danziger (1994: 296) has pointed out, throughout the twentieth century much institutionalized, mainstream psychological research gradually found itself in agreement with the late-capitalist image of an "independent individual for whose encapsulated qualities all social relations are external." What we begin to see in the extracts quoted earlier, however, is that just as genetic and brain-imaging technologies emerged that made it possible to get some measure of the organic substrates of these individualized phenomena, so it has become painfully apparent that looking at a diagnosis, or a brain, or even a lone synaptic connection, independent of the social context in which has any of them emerged, may not actually be adequate for describing many categories of psychopathology. In other words, at the beginning of the twenty-first century, and even within self-consciously biological and reductive approaches, it seems now increasingly difficult *not* to bring the social back in.

This recognition is embroiled in a number of important recent developments, and I can only roughly sketch them here (see Meloni 2014 for a longer account). For example, critical in the postgenomic era, have been the dissolution of the *gene-for* paradigm, the growing concern with gene-environment interactions, and, in particular, the emergence of an epigenetic understanding of psychiatric and psychological problems, which has involved researchers "coming to accept that DNA sequences alone do not comprise the master plan of organic existence" (Rose 2007: 47; cf. Rutter 2005). Epigenetics describes heritable changes in gene expression caused by something other than DNA—sometimes including such environmental factors as parental grooming of an infant (Weaver et al. 2004). Without getting into significant detail, I want to draw attention to the sense in which such an understanding must enroll the social environment in discussions of psychiatric and psychological pathogenesis. As the anthropologist and STS scholar Jörg Niewöhner (2011: 285) has shown, "epigenetics forces biologists to think about genomes in context . . . context here is not understood within a reductionist mode of thinking that reduces other levels of analysis to feeding into the baseline of DNA sequence. Rather the approach is systemic focusing on the multiple interactions between different levels of analysis."

The psychiatrist Steven Hyman, among others, has written about one of the most famous examples of this process—the effects of childhood adversity on the way that genes are expressed in adult behavior. Looking at a recent study that suggested a relationship between childhood grooming and later responses to stress in rat pups, Hyman (2009: 241, my emphasis) argued that "the frontier [in psychopathology] lies in understanding the mechanisms by which environmental factors (whether experiential, metabolic, microbiological or pharmacologic) interact with the genome to influence brain development and to produce diverse forms of neural plasticity over the lifetime . . . the *experience* of rats is transduced into long-lived *molecular adaptations* that influence adult behaviour." Of course there were other varieties of twentieth-century psychology and psychiatry to which this would not come as news. But my interest is in the inescapability of such understandings now, within a vision of psychology and psychiatry that, if it wouldn't be fair to call it biologically reductive, has at least aligned itself with a quantitative, physics-based "big science" of brain scans and biomarkers.

A similar conceptual move can be seen in gene-environment interaction research, such as the now famous studies of Avshalom Caspi, Terrie Moffitt, and their colleagues (Caspi et al. 2002; Caspi and Moffitt 2006), which have spun social context quite directly into the molecular structure of psychiatric distress. More than ten years ago now, Caspi and Moffitt (2006: 583) claimed that "the gene-environment interaction approach assumes that *environmental pathogens* cause disorder and that genes influence susceptibility to pathogens." Thus there has been a self-conscious move toward thinking about the effect of social life, not only within the remnants of psychoanalysis and social psychiatry, but also within specifically neuroscientific and genetic research on psychopathology—now facing a "witches' brew" of biology and environment, as Michael Arribas-Ayllon, Andrew Bartlett, and Katie Featherstone (2010) have put it. But I also want to stress that what makes up this brew is not (or at least not always) some thin, impoverished view of immediate environmental inputs. The "social" at stake in this space is, at least sometimes, the social as most sociologists would understand the term—namely, a series of complex, and historically situated, structuring webs of social inter-action, culture, and meaning.[5]

These still relatively recent understandings are deeply present, in all sorts of ways, in my discussions with biologically focused, big science–minded neuropsychologists and neuropsychiatrists. Consider this interview extract, which I quote at some length, because I think it well illustrates the complexity with which these scientists are grappling. It comes from someone who had been involved in funding a project to see how far back in infancy an autism diagnosis might be pushed. The answer, as it turned out, was: not so early. She said:

> One of the things that's coming out of the [research project]. . . . This is kind of a relatively new finding, the first thing they found they're track-ing these children over time, was that they expected them to fall into one of two groups, because they're doing similar work with children who are at risk of autism, so they had some notion of what their control group looks like—and they thought that their baby siblings would either look like the control children, or they would look as though they they're going to head down the autistic route. But what they found, as a group, they sat some-where in the middle. They were neither like the control children, nor did they display early signs of autism, but they did develop differently to other children, to control-group children, to typically-developing children. And then what they're finding now is that some people sort of start to head off towards the autistic side, and then veer back again. Um, and they obviously carry, in many cases, the same genetic risk as the sibling that has already got a diagnosis of autism—so some sort of protective factors are coming into play, we don't know what they are, but it's actually—the original notion was that you would be able to diagnose autism much earlier, and you'd be able to start intervening, and so on. It's now starting to look as though, actually, three [years old] is it.

In other words, as I understand it, this project has shown the limits of what can be deduced from biological information alone: kids with the same genetic load don't just take different paths; they even veer off these paths and then sometimes come back again. And no one really knows why. As another researcher put it, it's "a dynamic process, and that dynamic process is really underconsidered in developmental research generally, and particularly when you think about developmental disorders." What this clearly indicates is the degree to which, within the past few years, what Hyman (2009) has called "the frontier" of neuropsychological and neuropsychiatric research has shifted from being the province of a basically organic and biological science to a practice that must now, and without great training, begin to take account of shifting movements between biology and things like experience, individual biography, family relationships, social context, and so on. As anthropologist Rayna Rapp (2010: 669) has put it, "an appreciation of complexity and nondeterminism" within developmental psychology "has replaced an older enthusiasm for the deterministic one-way rules."

I am not at all arguing that there is anything unscientific about an attempt to think the relationships between, for example, brain development and family context. Nor do I claim that scientific rigour, reduction, and technique are defeated by a recognition of the generative role of social relationships within the formation and experience of mental distress. The literature discussed above does a good job of drawing rigorously scientific data from the psychogenic swirl of the social, the familial, and the contextual. But most of my interviewees—neuroscientists as they may well be—are also enrolled in a rather longer historical narrative, usefully described for us by, by Chris Frith (2007), as (I paraphrase) the defeat of the softly subjective by the bigly scientific.

As it becomes increasingly apparent that this transition is more complex than first suspected, these claims to science begin to look rather shaky. Within my interviewees' intellectual careers, a rigorously scientific approach to the kinds of things they were researching had gone quite concretely into gene-sequencing and brain-imaging laboratories and into the isolated body of the individuals concerned. But it had then (surely in quite an unexpected development) found itself having to *once again* account for the environment, and context, and social life, and things that were emergent, and fluffy, and maybe even a bit arm-wavy. Suddenly, entirely respectable neurobiological review papers, from major authorities on developmental disorders, were talking about racial discrimination and local authority housing policy (see note 5 in this chapter).

It seems to me that with these invocations of *science* within my interviews, and the very clear and unambiguous demarcations drawn between my interviewees' own work and some enchanted premodern past, some boundary-work is happening. As the neurobiological underpinnings of psychopathology increasingly get dragged across disciplinary and epistemic borders—fMRI studies on the one hand; analyses of racism in housing policy on the other—so does it become more necessary for practitioners to think about where exactly their own work is situated. This is not because they have a crude or reductive view of what gets to count as science—it's because they are tracing the neurobiology of autism across the border between the big science commitments of an imaging neuroscience and the awkward, hard-to-read sociality of so many psychological and psychiatric diagnoses. The insistence on "science," and not Freud or "faff," within these interviews is one way of picking a path through these subtly shifting borderlands; it helps these researchers to move between psychological practices that may be safely called *scientific* and a science of psychopathology that has to account for the machinations of history, society, and politics all the same.

COMPROMISING POSITIONS

My suspicion is that any project that interviewed roughly the same numbers of psychological researchers as this would find some element of the same conundrum. But I suggest, nonetheless, that the presence of autism is not incidental here. I have long been interested in the place of the social in accounts of autism: through case descriptions (Sacks 1995), diagnostic manuals (APA 2013), familial accounts (Park 1982 [1967]) and autistic autobiographies (Grandin 2005), it often seems impossible to talk about autism, as either a diagnosis or an experience, without *also* talking about the specific understandings, meanings, and sensations that mediate some person's social environment. A lecturer told me:

> Autism's not just about brain development per se. Obviously the environment affects brain development, and we need to figure out what environmental factors might actually impinge on children's development, in a positive way. And one of those might actually be social interaction in the classroom, or outside the classroom. And there are other things, like family . . . family structure. There's lots of studies in typical kids showing

that . . . um, having a sibling improves theory-of-mind skills. Because you just interact with the sibling, and you talk more, and you talk about other minds, kind of thing. So does it do the same for kids with autism? Or, what kind of . . . those are important things to answer.

Here we begin to get a sense of the tricky back-and-forth between a brain-based account of autism and the ongoing impact of the social environment—both on what that autism eventually looks or feels like and on whether or not it is even diagnosable. Later, this lecturer said:

> The possible focus on biology is that it's deterministic, and I think we have to be careful about that—because it's not deterministic [. . .] there's not one path, you know, if you've got autism, you don't necessarily go, in fact we have no idea which path you'll necessarily go down. It's not just the case . . . there are ways in which the environment can modify one's autism. I just don't think we understand what those conditions are, at the moment.

It is true, of course, that something very similar might be said about many psychological and psychiatric diagnoses, particularly about the subset of those diagnoses explicitly recognized as developmental. (We might also recognize that the social is very much at stake in, for example, schizophrenia; see Littlewood and Lipsedge 1997). But autism has a long tradition of discursive and clinical references to the presence, nature, and meaning of the social as well as to a kind of autistic *culture* or *planet* that may have its own account of these qualities. If claims to science in psychology are attempts to negotiate that discipline's troubled epistemic boundaries—and in particular the incursion of the social into the genesis of psychopathology—then studies of autism may be especially potent spaces for this negotiation.

First, we can say that perhaps more than any other diagnosis, *the social* has always been what is actually in question in autism—that it has historically been thought of as a disorder, more than anything else, of social interaction. The early descriptions of Leo Kanner, who published the first clinical account of autism, are the obvious guide here. Describing his very first patient, Kanner (1968 [1943]: 218) noted that "it was observed at an early time that [Donald] was happiest when left alone, almost never cried to go with his mother, did not seem to notice his father's homecomings, and was indifferent to visiting relatives." Kanner's second case, Frederick W., was described as having "always been self-sufficient": Kanner quotes Frederick's mother: "I could leave him alone and he'd entertain himself very happily, walking around, singing . . . when we had guests he just wouldn't pay any attention. He looked curiously at small children, and then would go off all alone. He acted as if people weren't there at all, even with his own grandparents" (ibid.: 222–23).

Read through the prism of the twenty-first century, Kanner's shrewd and self-aware article reads like a ghostly collection of autistic representation through the ages: the busy father, the college graduate mother, the strange eating habits, the gifts for memory, the regressions, the false dawns, the silences—so it goes on. But above all, what united these children, what convinced Kanner that he was dealing with a phenomenon that "differs . . . markedly and uniquely from anything reported so far," was the preponderance of a specifically *social* alterity. "The outstanding, 'pathognomonic,' fundamental disorder," Kanner (ibid.: 242, emphases in the original) concludes, "is the children's *inability to relate themselves* in the ordinary way to people and situations from the beginning of life . . . there is, from the start, an *extreme autistic aloneness* that, whenever possible, disregards, ignores, shuts out anything that comes to the child from outside." Amid the diagnostic and etiological fluctuations of the decades that followed, this cardinal feature never left autism. For a category principally famous for its diagnostic and phenotypic heterogeneity, this consistency is really remarkable. Although "several other developmental disabilities typically accompany autism's social dysfunction," notes psychologist Ami Klin (2002: 895), "the core social disorder defines the condition." To think about autism has always been to think about the social: autism research has long been a boundary point for the frontiers of psychology and context, of individual biology and nurturing environment.

But this relationship can be inflected in a different way, because autism has also long been a testing ground for exploring what gets to count as social in the first place. This reflects the degree to which, within autism research, we find a notably cosmopolitan discussion of sociality-in-the-first-place. Here, I am thinking, for example, of long-standing claims to an autistic culture. I am not only referring to such well-known claims as social scientist Judy Singer's (1999) smart deployment of the word *neurodiversity* or

Amanda Baggs's (2007) complex elaboration of the subtly sensed and tactile grammar of her social world (Baggs and Singer are both themselves on the autism spectrum). I am also referring to less obviously self-conscious accounts that manage to unfold the same sense of a distinct autistic cultural presence. I am thinking, for instance, of Oliver Sacks's well-known encounter with the autistic scientist and writer Temple Grandin, documented in his 1995 book *An Anthropologist on Mars*—the striking title of which comes from Grandin's description of herself trying to make her way in an *alien* culture. That same sense of autism-as-cultural-difference is echoed in STS scholar Chloe Silverman's (2008: 325) description of social-scientific writings about autism as comprising "fieldwork on another planet" in the title of science writer Steve Silberman's 2015 book about autism, *Neurotribes*, and philosopher Ian Hacking's (2006b: 3) identification of the specific difficulty faced by parents raising an autistic child—namely, "your child is an alien."

In the opening chapter of her memoir, autistic primatologist Dawn Prince-Hughes (2005: 11–15) vividly describes herself as inhabiting a "culture of one"—a culture that, she makes clear, is much more attuned to the thrills, pains, and rituals of gorilla life than it is to the social niceties of zookeepers and their small talk. Or as the autistic author Lucy Blackman (2005: 149) has similarly put it: "It may be that the social deficits which are the cornerstone of an autism spectrum diagnosis tell us far more about the person who made them markers for such a diagnosis than about the child whom he observes. I realise that social life and affections are essential for being human, but I still wonder whether the 'me' factor is properly understood. That is, the whole testing procedure is somehow actually constructed on whether the tester observed the person to socialise in a way the tester understood to be socialisation." What these descriptions show is that not only has autism long been a disorder of the social, but, as interpreters of both a diagnosis and a lived experience, autistic people have consistently challenged those who would do research with them to *rethink* both what gets called social and where it can be sought (cf. Davidson 2008; Ochs and Solomon 2010).

This sense of autism's particularity for psychological research, and the degree to which that particularity might contribute to the story that I am trying to tell in this book, was brought home by another interviewee's use of an analogy with ADHD and the single-gene disorder phenylketonuria (PKU). She said:

So PKU, which I just mentioned, this genetic disorder, it tends to be thought
of as a biological disorder, because it has a single biological mutation, and
everybody who has the disorder has that single mutation. Whereas ADHD
is generally thought of as a behavioral disorder, because everybody admits
that this is just a group of people who have been gathered together, who all
have sort of some attention problems and some hyperactivity problems, but
there's basically nothing that draws those kids together at the cognitive or
biological levels . . . Autism tends to be the one that falls in the middle.

It is precisely this sense of autism research as something that "falls in the
middle" that I am getting at. It is in the context of such a falling, of such
a sense of betweenness, that my interviewees experience a desire to more
carefully (and even assertively) *feel* the boundary between science and the
social. It is not only the case that these otherwise nuanced neuropsychologists want to be taken seriously as hard, reductive, objective scientists. The
point is that the disciplinary and epistemic border crossings of the neurobiology of autism *require attention to these limits*. The invocations that I
discussed earlier are not simple, self-interested claims to the status of science. They are expressions of the ways that neurobiological research on
autism requires hard attention to the ongoing labor of social and scientific
bordering.

I want to concretize this claim, and close the chapter, with one final
extract from my interviews. It comes from someone who was more expert
on the complexity and valence of these things than most, being both a
researcher on autism and the parent of an autistic child. Where this extract
picks the conversation up, we had been talking about psychoanalysis and
its infamous "parent-blaming" history in autism.[6] I brought up a mouse
study that I had been recently reading (Mines et al. 2010). In a way that I
did not quite understand, it seemed to me the study was in some ways a
return to thinking about parents in autism. She said:

The parent-child is a very sexy subject at the moment, and you know it
tries to sort of steer a clear course, but it's difficult. It's a tricky one. It's a
tricky one. . . . It's a fine line, I think, but it's interesting that it's coming
back to that.

I asked her how she felt about this return, as both a researcher and a parent. "Um . . . ," she said. After a long pause, she laughed and continued:

> I have issues with it, I do have slight issues. And, you know, it's interest-
> ing—I've sat in meetings where we've discussed genes and, you know,
> where I've looked around the table and, you know, people have been saying,
> "Well, you know, and clearly it's not surprising that parents would have
> an input because the parents themselves are probably slightly autistic, or
> at least broader autism phenotype kind of thing." Um, and I've kind of sat
> there at the table thinking, "Actually, you know, I think I'm probably more
> socially adept than most people around this table" [*we both laugh*]. But, you
> know, I won't say anything. So I find myself being in compromising posi-
> tions, every now and then—surprisingly infrequently actually, given the
> position I have. But, um, I do find that quite . . . quite difficult, um . . . And
> because I, fundamentally, going to the original question, I do fundamentally
> believe that it is a neurological thing. And, you know, even just the way
> that it happened with my son, there was clearly this kind of window, there
> was this point at which something is happening developmentally. Because
> he was apparently fine, and then something happened. And I don't believe
> it's because he had a vaccine, or anything like that—it was something, you
> know, there was a point in development at which something, you know,
> he went one way, and most children go another way, and, you know, I feel
> that's very clear.

Caught both personally and professionally between the parent-child relationship and "something happened," this scientist-parent exemplifies the
delicate crossings that characterize this space, including the rhetorical, practical, and affective labor required to position herself among them. It should
be no surprise that she expresses, with special acuity, the "compromising
positions" inhabited by so many neurobiological researchers on autism.
My argument is that what look like crude and simplistic statements about
science are better heard as early ways to think and talk about such com-
promises. They might be heard as a way of moving between the ongoing,
generative loops of familial and social life, on the one hand, and the hard
science of individualized brain biology, on the other. This complexity, for

the autism neuroscientist, is characterized, by the "veering back" of genetic and environmental inheritance within which all human action has to be interpreted. But it is also well represented by the long-standing and richly *social* nature of autism, which not only names a disorder of interaction and communication but also marks a demand for particular forms of sociality to be recognized and valorized in the first place. As this last interviewee points out, emerging forms of research are actually rearticulating, in new forms, border crossings that have surrounded autism research for many decades. They are doing so precisely at the moment when many researchers thought they might be able, finally, to move on.

TAXONOMIES

The truth, of course, is that neither psychology nor psychiatry has ever been unambiguously associated with the natural sciences. Or at least, to the extent that either can now legitimately make such a claim, this has only been a recent development (Rose 1985, 1996a; Luhrmann 2001). The strong temptation is to write another denunciation of psychologists' or psychiatrists' attempts to ward off the social and cultural elements of their research objects. But if this book is about anything, it is about the impossibility of getting any meaningful analytic purchase on this area if we enact the kind of rigid cut that such an argument would demand. In fact, this is not a story about externality to science; instead, it is a story about the shifting borders and taxonomies of what gets called *scientific* within this century's study of psychopathology; it is a story about the ontological complexity of the world that is being named and organized under the rubric of such a study; it is a story about some unstable intellectual and disciplinary spaces that seem to demand more fluid and open relations (in practice, if not always articulated as such) to epistemology, disciplinarity, science, and the social.

One of the main things I learned from being around autism neuroscientists is the delicacy with which an unequivocally scientific and given object might *still* be traceable within an unstable and contingent world of social interaction, communication, politics, and so on. I am describing here my own realization that these scientistic moments are not the desperate pulling down of a veil to cover psychology's inadequacy; nor are they moments of psychologists' blindness to their own shortcomings. What they may actually amount to, in the end, is a window onto a particular moment in the history of psychology—a moment in which a scientific account of psychology's object has, for reasons that are complex and multifaceted, but without much warning in any case, become strangely allied to a methodological and conceptual entanglement with the social.

I cannot stress enough: it is not at all my point to deconstruct the scientific claims of a contemporary psychology, or to say that, whatever its pretensions to the contrary, psychology is still not a real science. The strange thing is that the opposite is probably true. The simultaneity of contemporary psychology's claim to science and its investment in the social are not problems for, but in fact elements of, one another. I don't think it's simply that we don't understand this in the social sciences (although we don't). I think we are actually more broadly in danger of totally missing a moment of striking openness in the practice of neuropsychology. In other words, it seems to be that case that the coming-into-science of psychology has not only not put up a barrier against the social. It seems to have had—and this, I guess, to everyone's surprise—precisely the opposite effect.

5

SEEING THE UNICORN

I HAVE BEEN INTERESTED, SO FAR, IN THE TENSIONS, DIFFERENCES, contradictions, and ambiguities that structure a neurobiological account of autism. But I have also been keen (maybe too keen) to stress that I have no interest in mobilizing this account in the service of an argument about (1) whether autism (or any other neurodevelopmental diagnosis) is really a brain disorder, or else some kind of social construction; or (2) whether neuroscience (or any other biological science) truly gets at the organic substrate of the phenomena it investigates, or if it only reproduces the cultural assumptions of the day and so on. At the heart of this book is an attempt to set these kinds of binaries to one side and to draw into view the strange circulations of social and biological material that are set in motion by the neurobiological study of autism.

One of the reasons that I have been so interested in this circulation is that it interrupts any attempt to install a polarity of ambiguity and certainty in experimental practice: following the work of Karen Barad (2007), I have described a scientific endeavor in which we begin with relations, interactions, interpretations, and ambiguities. We start there not to destabilize a scientific endeavor, but with precisely the opposite intention: to gather together the components from which a stable scientific object might actually come together. To put this more baldly than I would maybe like: the multiplicities and ambiguities that I have described, up to now, are not signs of the increasing *instability* of autism as a neuroscientific object; they are actually methods for thinking about how a more or less coherent autism gets traced together by the objects and agencies that make up the new brain sciences.

FOLLOWING THE TRACE

None of this comes from me. In fact, the impetus for this claim comes from two sets of metaphors that came up again and again during my interviews. One was a constant reference to autism as a dispersed and disaggregated phenomenon—something that can only be understood from multiple dimensions or via very different levels of understanding (genes, cognition, behavior, brain anatomy, somatosensory thresholds, time, the environment, life experience), which do not necessarily connect with each other in an obvious or straightforward way. The other was a way of talking about neuroscientific research as a practice of *shuffling, connecting, assembling, tangling,* or—and this I take as the guiding image for the book as a whole—*tracing.* In this final chapter, I think a bit more carefully about these metaphors. At stake in them is an empirical object that is a bit unfashionable but that I find myself drawn to all the same—and this is the scientific fact.

For Bruno Latour (2008: 39, emphasis in the original), if we are accustomed to thinking of scientific facts—like the neurological basis of autism spectrum disorders—as things that are *indisputable, obstinate, simply there,* we would do well to instead move to thinking these *matters of fact* as more *matters of concern,* as things that need to be *"liked,* appreciated, tasted, experimented upon, mounted, prepared, put to the test." It's important to note that the shift to "concern" is not a repudiation of the obstinacy of fact—Latour makes this clear in a much cited mea culpa (Latour 2004)—but is rather a way of empirically describing how facts come to be facts. "The question," Latour (ibid.: 231, emphasis in the original) points out, "was never to get *away* from facts but *closer* to them, not fighting empiricism but, on the contrary, renewing empiricism."

Donna Haraway, no doubt rightly refusing to share Latour's postmillennial guilt, is still happy to talk about "matters of fact." But for Haraway (1997: 267) the fact has *always* been instantiated as "a crucial point of contingent stability for possible sociotechnical orders, attested by collective, networked, situated practices of witnessing." In place of what Latour has called a "second empiricism" then, Haraway (ibid.: 267, 11, 120) offers "contingent stability" and "figural realism"—images that hold together a framework of practice in which the relational "constructedness" of technoscientific hybrids is "not in opposition to their reality," rather being "the condition of their reality" and even "fast becoming the sign of reality as such."

This chapter explores the emerging facticity of the neurobiological account of autism, as it, too, is brined by logics of collectivity, network, relation, and appreciation. I lean on Haraway and Latour in my insistence that identifying an autism that exists at different levels, which don't easily relate to one another, and then thinking with a neuroscientific practice that works to trace a scientific object *across* these levels, is not to diminish either autism or neuroscience. By the end—and perhaps I will fail in this endeavor—I want to have specified how it is, precisely, that neuroscientists participate in drawing scientific agencies through the different levels of their appearance. This means expanding on what I have up to now described as a tracing neuroscience. While I remain close to the data, it is crucial that the experimental logic I call *tracing* is understood as a way of thinking about neuroscientific research more generally. My hope is this: thinking through traces will help those of us in the social study of the life sciences who want to understand what's happening in a space like this one, to let go of the critical suspicion that (still) remains the go-to interpretive framework for understanding scientific ambiguity—as well as being the leitmotif of so much of what goes on under the sign of "science and technology studies" more generally. At the heart of the chapter, I ask: What might it mean to follow the trace?

PEOPLE ARE JUST TOO COMPLEX

The background to this interest comes—as does so much of what I write in this book—from a jejune early set of questions that I put to neuroscientists (of course I stopped as soon as I knew better), in this case about what they took to be the *cause* or *causes* of autism. Authoritative accounts of the cause of autism, especially those written for nonscientists like me, are often quite opaque and generally say that causes are currently not known (but are likely to lie in some combination of biological and environmental factors).[1] This noncommittal account is unsurprising, given the cultural side effects of claims that were at least interpreted as having some causal weight (such as those ventured by Bettelheim [1967] or Wakefield et al. [1998]—the latter since retracted). This all made me interested in how autism neuroscientists would talk about cause in an interview and, in particular, how they would negotiate the awkward trajectories of the environment and the body within popular histories of autism causality.

But when I brought it up, my questions were not answered with accounts of how cause was currently known or not known, or with arguments either for or against biological and environmental cause. Instead, discussions tended to circle around a view that cause would only ever be known by picking autism out across the multiple levels on which it existed, or by working through the different levels of understanding within which it had to be appreciated. The point was that there would likely never be a single cause found for autism, but that it might possibly (one day) be shown how a big-enough subset of numerous, independent, contributing factors, across the levels of genes, environment, behavior, anatomy and time (among others), would sometimes, for some subset of individuals, and for different reasons, congeal into a set of characteristics that we might reliably call *autism*. A PhD student said:

> I don't think it's going to happen that we will find a single cause of autism, and I think research should . . . . it shouldn't necessarily give up on trying to find a cause, at least a single cause, I mean finding, sort of, numerous different contributing factors is definitely useful, and I think it's definitely going to be more the case that there's lots of different contributing factors, which kind of come together, and that sort of manifests as autistic spectrum, or something like that.

It's interesting that the student attempts to be generous to specific causal accounts of autism, and she lends vague support to the idea that other researchers "shouldn't necessarily give up on trying." But even in the midst of trying to speak up for the search for "a single cause," she begins talking about "numerous different contributing factors, which kind of come together"—which is something very different.

Even this loose idea of things "coming together" was, for others, a bit suspect. One person said to me, during a similar conversation: "I mean it's terribly easy to think about this neat causal chain. But I don't think it's like that. And I think it's different in various cases." Another said:

> [A final causal explanation of autism is] probably going to be a sort of a multilevel thing—so genetically you'll be able to describe the different gene

variants that can contribute to it, but to me the final causal pathway, if you want to use that sort of language, is most likely to be a neurodevelopmental story [. . .] we already know there are a lot of factors that can lead to this causal change, so genetic, environmental and so forth, so it's not going to be a simple causal story—you will have multiple different possible factors going into one final common pathway which will be a neurodevelopmental story, and then that'll have multiple widespread consequences subsequently. That's my guess.

I am initially struck by this researcher's reluctance even "to use that sort of language." Perhaps even more striking is the reference to autism as a "multilevel thing," centered on the different collections of genetic variants, neurodevelopmental pathways, and variable consequences in the manifestation of symptoms. This is exactly the sort of account that oriented me to the sheer multiplicity of autism—which is not only to say that it is visible on different levels (which is trivially true for likely any biological disorder), but that there is little sense of autism's path-dependence *across* these levels.

In other words, it's not only that autism is both genetic and neurodevelopmental, but that a particular subset of genetic factors for one person's autism may have nothing to do with the neurodevelopmental unfolding of autism for another (because the neurodevelopmental path is linked to a different subset of genetic factors), and these may *both* be independent of the symptoms that are actually diagnosable as autism in a third (because, of course, those symptoms might be traced to a whole other subset of autism-related genetic mutations and to a quite different neurodevelopmental pathway). This kind of reflection drew my attention not only to the well-known variability or *heterogeneity* of autism but, more significantly, to the difficulty of working with any coherent sense of autism that might, even as a concept or a model, be held together across this all-pervasive sense of difference. Consider this account, which came from a postdoc with a background more in neuroscience and neuroscientific methods than in autism research as such. He said:

If you try and differentiate often between people with autism on a univariate measure, so I just take, like, somatosensory thresholds, right, on one test, then, you know, it's difficult to see what's going on there. [. . .] [T]he best way probably to think about autism is not . . . you're not going to find a brain locus for it, it's not going to be an "Aha! Right, there's this big problem here, and that's it. And we're just going to sort that out." It's: "Something happens in early development that causes multiple [*tape cuts out for a few seconds*] . . . , and we're dealing with the effects of that, and it might go in different ways with different people and we're not sure why yet, essentially." But at least all of those differences are different enough and consistently like one another enough, even though you might have to go into n-dimensional space to see it, that you can distinguish it.

His critical point is that using just one measure, autism remains indistinct—that it is difficult to find "an autism" using any single measure, such as an EEG measure of the somatosensory system.[2] But measuring different aspects of the disorder, in different people, across multiple levels ("n-dimensional space"), might give you—and this perhaps is the most that can be hoped for—"differences [that] are different enough and consistently like one another enough," such that something that looks like autism might be broadly isolated, at least as a kind of statistical aggregate suspended across these multiple levels. This is an argument that autism is not simply *heterogeneous* but that it consists *in* measures of statistical commonality between the various sets of difference, until something that looks enough like autism begins, even if only in terms of some statistically significant relation, to hang together.

Of course, this way of thinking about cause is not unique to autism. It is the sign rather of a more general and multidimensional "probabilistic" thought style emerging in neuroscientific approaches to psychiatric and psychological diagnosis (Singh and Rose 2009). But autism does nonetheless have a particular relationship to the way that cause can be modeled across different layers (in particular, the genetic, cognitive, and behavioral layers). For example, several of my interviewees referred me to John Morton's (2004) "causal modelling" approach to developmental disorders. As they described it, this model distinguishes so-called A-shaped disorders from V-shaped disorders, depending on where you see unity across cases. A-shaped disorders have widely differentiated genetic inputs (the bottom of the A), but behavior remains mostly stable (the narrow point at the top of the A; ADHD was one example provided by an interviewee). V-shaped disorders show the inverse pattern—a discrete and well-described genetic cause at the bottom, but with wide disparity in behavioral symptoms at the top (PKU was given as an example). What's interesting about autism, though, is that it shows an X-shape, which indicates disparity at *both* the genetic and behavioral levels but with some consistency at the level of cognition (the center of the X), albeit this is something that you can't really see or directly measure.

This goes some way to showing the degree of multiplicity that, for these interviewees, characterizes autism in particular. It also shows us the lack of an obvious winnowing or a single path between the behavioral and biological levels. As one interviewee put it:

> Let's say you have a bank of genes that are the autism genes. But it's not just "If you have that gene, you have autism." And it's not even "If you have this particular combination of genes, you have autism." You've just got to have enough of a mix of them, with enough of them affected in a big enough way, for it to produce symptoms. [. . .] Cognition is difficult because you can't see it; it's this loop between biology and behavior. But you can't actually get at it [. . .] [tests for cognition] are always behavioral tests. If you do something that you obviously think of as biological, like brain imaging—[it's] to get at the cognitive level. There's no way of actually getting at the cognitive level, apart from through biology or behavior. It's an invisible thing. It's a concept, in that sense. So it has to have some form of reality at a biological level, in some form, at some stage. But we just don't know at what level that's at.

Two things are worth drawing attention to here. One is, again, the difficulty of connecting the biological and the behavioral: even if the behavioral level was consistent and well described in autism—and it is not—this tells you very little about what's happening at the genetic level. As noted, very different combinations of genes, with no commonality between them, may be predictive of "autism," depending on the equally nebulous effect of the environment that the person grows up in (Persico and Bourgeron 2006). Second, though, is this sense of invisibility around the cognitive picture of autism—"it's a concept, in that sense"—which may be the only level at which, according to this set of interviewees, the definition of autism is at all robust, but which still has to be inferred through large, disconnected, and differentiated pools of biology and behavior. This goes a long way toward describing what is hard about autism's multiple appearances across the different levels: it captures the difficulty of finding any path for autism between these levels—such that some now even wonder if autism is *only* a phenomenon of multiple levels.

Here I have limited myself to a discussion of cause, but the idea that autism was a phenomenon of different levels of understanding, and that these did not always or obviously come together into a coherent disorder, was a feature of several other parts of my interviews. I saw this way of thinking, for instance, in discussions about translating between laboratory and clinic: as one person said, "There's this disconnect, because the cognition's got nothing to do with the way clinicians look at it and how they're diagnosing it, but that's how *we're* looking at it." I also saw it when people talked about the potential for treatment. The interviewee quoted at length above went on:

> I'm never going to know, never in my lifetime am I going to know how all of
> the different possible factors that influence a person, or are likely to affect a
> course of treatment or a behavioral plan, or whatever it is, for that person.
> Because people are just too complex. There are just too many factors. And,
> let alone working out how they're going to affect an individual, we're prob-
> ably not even going to know what all those factors are.

Others talked about the difficulty of maintaining a single autism across the lifespan of an individual, or within a cohort of research participants. One researcher said: "Probably a lot of the guys I've got are probably more on the Aspies level."[3] After a pause, he continued: "Which doesn't exist anymore, according to the *DSM*."

The question that emerges, of course, is whether autism actually coheres at all. There are a few different answers to this. One is to say that autism doesn't necessarily hang together, in the way it is sometimes imagined—and we can make sociological (Nadesan 2005) and neuropsychological (Happé and Ronald 2008) arguments to broadly that effect. Another answer, and I touched on this earlier, would be to say that this might just be a sign of a broader psychiatric thought style, in which diagnoses are defined according to their probabilistic relationship to one another *across* levels. A third argument would be related to the one that I attributed to several interviewees in chapter 1—a more pragmatic claim that autism is a "diagnostic category" or a "symptom checklist," and anything else is, for now anyway, empty speculation.

I want to venture another way of thinking about this, however. What if working toward autism's coherence is a potentially useful way of characterizing what autism neuroscience *does*? What if being traced together, within a sociotechnical order that includes the new brain sciences, the genomic sciences, the environment, cognition, and a whole host of other things is a way of thinking about what or how autism *is*? Can we hear, in these accounts, an emerging sense of autism's "contingent stability"—a way of talking about the degree to which autism is indomitably "present" in these spaces (Murray 2008: 16), but in which that presence is nonetheless *figured* (I use Haraway's term) by a complex and entangled neuropsychological research practice? Let me unpack this by focusing on the last bit first, the strange and knotted nature of the contemporary brain sciences.

A MIX OF EVERYTHING

What is actually interesting, here, is not the preponderance of differences between levels of understanding but actually the way that these levels are (perhaps slowly sometimes) drawn or held together by different researchers. This possibility was initially suggested by an interviewee who had experience in research management and who had worked at a senior level within a UK research funder. She said:

> I think one of the interesting things about autism research, as distinct from some other fields of research is the degree of commonality of view across—I mean, they'll tell you the balance is wrong. You know, the psychologists will tell you we're spending far too much time doing genetics and brains scans, they'll tell you, you know, that will only take you so far. And on the other hand, you've got the neuroscientists saying, "Well, the psychologists haven't produced anything that helped people with autism for twenty-five years" [*laughs*]. [. . .] But I think generally [the scientists'] view of what autism is, how it develops, what the issues in autism are, how it translates into social difficulties, you know, the things you might do to alleviate that situation, I think there's a lot of common ground between the different scientific disciplines.

In this section I don't want to focus on the degree of difference. Instead, I want to go with this perhaps less critically exciting account of the persistence of "the commonality of views across" and of the establishment of "common ground between the different . . . disciplines." Laughing a bit at my description of the purpose of her organization as "connecting things," this interviewee later said: "I think acting as facilitators, brokers, harriers—I think that's an effective role that a small [organization] can play." Consider also the following quote, which comes from a senior researcher who maintained an identity in both the clinic and the laboratory. He said:

> There's so much science that's relevant to understanding a complex brain condition like autism, a developmental condition as well—both in regard to brain development, and also to development throughout the lifespan.
> [. . .] But I'm the sort of person who wants to know a little bit about all the sort of levels of understanding. So I try and keep up at least at a basic level with what the sort of genetics story is, and that's interesting because for the first time in quite a long time it's changed in the last few years, and opened up potentially. And then thinking about sort of brain development, and some of those issues around how the social mind sort of develops and gets put together is a really interesting story, potentially, and I've been thinking about how that fits onto the emergence of autism.

Here, what's interesting is not only that there are "all the sort of levels of understanding," but that the different levels are all still relevant for him. "Autism research" cannot be a singular thing; it is, at the same time, a "genetics story," something to do with "sort of brain development" and also a question of "how the social mind sort of develops." But the key thing is that he "wants to know a little bit about all" of these.

I am trying to get across a sense of a research practice that wants to attend to, precisely, "all the sorts of levels," which is interested in "the commonality of views across"—and which is therefore not entirely averse to diagnostic objects that drift through lots of different kinds of manifestations and understandings. As another person put it:

> Behavior always has brain correlates—there's no such thing as free-floating behavior. Which doesn't mean to say I want to go down a reductionist route.

> If you can understand what those correlates are, then it is at least an inter-
> mediate step toward . . . possibly physical-based interventions and, further
> down the line when you get to first causes or etiology, possibly to prevention
> [. . .] I see a causal tangle. So I'm not going from etiology to neurobiology to
> behavior in any neat way. Because they all feed backwards and forward.

Note, in particular, the relationship between the way that this person thinks
about the multiplicity and variety of autism ("I see a causal tangle"), to how
they conceive of their research practice as exactly that which can "feed back-
wards and forward—albeit not in "any neat way"—in order to work, slowly
and differentially, *through* the many layers of this tangling.

References to this kind of tricky, back-and-forth motion came up again
and again: "As far as I see it," said one interviewee, "the behavior needs to
inform how we approach looking at anatomy, that's what I do." Another
person spoke about her delight, as a PhD student in neuroscience, upon dis-
covering "these multiple strands of evidence that would show you how your
model was apparently correct or where it fell down." And later, talking about
the likelihood of a big breakthrough in autism biomarker research, she said:

> I would probably tend toward thinking [there's not going to be] such a big
> breakthrough if it's not set up as a sort of massive long-term study, or set
> of studies, with a lot of different experts from different areas, feeding in
> kind of the state of the art . . . I tend to think, "Oh, another study that looks
> at the genotype," for example. Well, they come up with different kinds of
> candidate genes, and why would anyone think that is going to find the true
> candidate in isolation from other things?

Here I am drawing attention to the unwillingness among my interviewees
to pursue a research practice that stands "in isolation from other things," an
unwillingness that enables them to appreciate the "tangle" of all of autism's
"levels of understanding." This requires not only an ecumenical awareness
of other views, but the active pursuit of "commonality," by moving "back-
wards and forwards" across the different epistemological and disciplinary
layers: neuroanatomical, behavioral, cognitive, genetic, environmental,
experiential, and so on. I found this kind of self-positioning to be a central

feature of many of my interviewees' descriptions of what was involved in thinking neuroscientifically.

In many ways, of course, moving across disciplinary and epistemological levels is precisely what the new brain sciences were set up to do. As the neurobiologist Steven Rose (2004: 3–4) has pointed out (I mentioned this quote in my introduction): "What were once disparate fields—anatomy, physiology, molecular biology, genetics and behaviour—are now all embraced within 'neurobiology.' But the ambitions of these transformed sciences have reached still further, into the historically disputed terrain between biology, psychology and philosophy: hence the all-embracing phrase: 'the neurosciences.' The plural is important." Indeed, as noted elsewhere, one of the most significant outcomes of the "decade of the brain" was the number of scientists—molecular biologists, computer scientists—not specifically trained in neurobiology who found it both possible and advantageous to begin describing themselves as "neuroscientists" particularly (Jones and Mendell 1999). In a related sense, my interest now is in the degree to which *living with* connections between different areas, even where no dominant narrative connects them, might be a core feature of the new brain sciences. As a postdoc put it to me when I asked what had led her to her current research:

> It's just that I was very interested in psychology, and I just wanted to find a way to get to working with people with cognitive impairments, and I was very interested in autism, so, uh, via this route I knew that I could work and do research on autism. And I really like the fact that cognitive science is a range of sciences actually. There's a computer aspect, there's a social aspect—it's a mix of everything. [. . .] I like this, um, diverse, um, this approach of mixing diverse ways together to investigate an aspect— especially as there's so much going on, and it's not just biological, there's a lot around the issue of why some children develop autism.

What I like about this account is her telling of the discovery (or perhaps her chiding of my assumption otherwise) that cognitive neuroscience might be something that did *not* run against her more wide-ranging, human interests (which we had just discussed). But not only is her research implicated in "a range of sciences," these include such different domains as computer science, biology, social research, and so on. To do neuroscience, for her, was already to have some kind of awareness of the different things that needed to be joined up.

It was the same for a neuroscience PhD student who spoke to me about her interest in the role of "social priming" in autism and the different ways that she might actually think about investigating it. She began by talking about the difficulties of looking for this phenomenon at the behavioral level and argued that it might be easier to think about it at the neurological level—for example, by thinking about connectivity:

> If you find that the connectivity is weaker, you can start asking questions like, "Well, why is that?" Is it because there are less connections, like physical connections, axons, between these areas of the brain and these areas of the brain? Or is it because those axons are narrower, less myelinated, something like that?" I just think that that [neuroscientific approach] gives you, you know, a lot of ways of exploring the problem. And then hopefully—so you should then really go back to development and try and find out, well, why is that? Is it something that's genetic, that's associated with just these people who are born this way, or is it due to experience? But when you get to the experience question, then I think obviously you need behavioral experiments as well, because you really need to take into account the experience that this person has had, and that's a behavior, so you need to use appropriate behavioral experiments for it.

I jumped in to say that I thought it was interesting how you could go from behavioral experiments to brain imaging, and then back out the other side, as it were—to something experiential or developmental. She said:

> Yeah, because if you ignore that and only look at the brain, then, then you kind of have nothing, because you just have . . . it's not nothing I suppose. [*Laughs*].

The very specific interest of neuroscience, for this neuroscience PhD student, was that it gave her "lots of ways" for thinking about a whole range of problems: Do people with autism have difficulties with social priming because they have fewer axons? Then you don't just need to scan brains; you also need to think about genetics and experience too. And if brain scanning is "not nothing" in the neurosciences, nor is it the only level at which questions can be approached. As another person put it quite bluntly:

> The basic, sort of, the bottom line, perhaps the most crucial insight [of the new brain sciences] is that our brains work at many different scales. So there are genetic scales. For instance, our neurons are the cells in the body that have the most prominent genetic expression of all cells in the body. So we have many scales there—anatomic scales, scales of genetics and gene expression. There is then this scale of the dynamic response—so, for instance, just in terms of time, expressing a gene before it is fully operational can take hours, or even a day or two. But obviously when we are talking to each other here, things are on a much more rapid timescale. So we have different technologies which basically probe into the different windows, the different areas, the different scales the brain is working at.

In interview after interview, neuroscientists consistently reported to me that to *do* neuroscience was to think at different scales—that the *work* of neuroscience was making sense of this scalar tangle. Another said: "Whenever anybody tries to set up some sort of dichotomy in brain sciences—whether it's nature versus nurture or anything else—initially it seems like a sensible question—rapidly people come and say it's neither one nor the other, it's a combination." This resistance of bifurcation in favor of a combinatory logic was a defining feature of my interviews with neuroscientists. But I am also trying to suggest that the presence of this logic has forced me to think more carefully not only about the inherent "multidisciplinarity" of the new brain sciences but about the kinds of complex, tangled, multilayered, and even incoherent research objects that such a science must be capable of thinking about and working on.

Also implied in this account, of course, is a particular kind of researcher—one who is both sensitive to, and collaborative with, particular kinds of objects. I am reminded, as I read back through these accounts, of Donna Haraway's "modest witness"—that complex and active scientific figure who is able to work through forms of "contingent stability" (Haraway 1997: 23, 267). In her description, the "modest witness" of the life sciences is no self-effacing Baconian "ventriloquist for the object world . . . endowed with the remarkable power to establish the facts" (ibid.: 24). Being particularly concerned with new sociotechnical orderings, and the strange sorts of object-stabilities that are worked across them, Haraway (ibid.) challenges us to recast both what counts as modesty and what counts as witnessing: "I would like to queer the elaborately constructed and defended confidence of this civic man of reason," she writes, "in order to enable a more corporeal, inflected, and optically dense, if less elegant, kind of modest witness to matters of fact to emerge in the worlds of technoscience."

Haraway (ibid.: 269) is especially attentive to the complex and ambivalent gendering of these practices and to their particular association with scientists who are embedded in what she has called "nonstandard positions." If Haraway is correct that "the exclusion of women and labouring men was instrumental to managing a critical boundary between watching and witnessing," then we must pay some attention to who precisely is at work in this space (ibid.: 33). Here I note the representation of women in my interview sample: as I mentioned in the introduction, and excluding those I interviewed from outside universities, I had a ratio of about two-to-one women to men in my interview sample. It is not my purpose to think about how this fact sits against a broader scholarship on the exclusion of women's scientific labor from laboratory spaces (Fox Keller 1977; H. Rose 1994). Nor can I do justice to the complicated and mostly US-based statistics on the place of women in psychology and neuroscience.[4] But if there is far too much going on in the long-running politics of these developments for such a short discussion as this, I still take seriously (amid discussion of a complex scientific practice of entanglement and combination), a reminder from Hilary Rose's (1994: 2), the sociologist of science, that "feminist biologists, in contesting the boundaries of nature and culture laid down by sociobiology, understood in a direct and practical way that as women we, our bodies and ourselves, are part of both nature and of culture."

I favor Haraway's conceptual apparatus because it draws attention to a possible relationship between the openness of these combinatory logics, the frankness of the modest-witnessing practices of (some parts of) the new brain sciences, and the presence of so many "nonstandard" bodies working in this space. This gets missed in the more self-consciously disembodied "actor-network" accounts that I discuss next. As an interviewee quoted earlier put it:

Going in to study cognitive science where there is a lot of computing, for
example, can be a bit, uh . . . it was hard for me, because I'm not really
interested in that. I knew that—and especially because I'm really interested
in rehabilitation—I knew that if I could master these computing skills, I
could help develop programs that could help children with autism to communicate better. [. . .] I need to master this aspect of science as well, even if
it's been painful. It was a conscious choice that I've made.

While fully acknowledging the complexity and bipolarity of the politics
that inhabits claims like these, and while trying not to mark "nonstandard"
accounts with a responsibility for care, I am mindful of the "feminist recon-
struction of rationality" that Hilary Rose (1994: 49–50) has argued for, "in
which senses of responsibility and caring are restored within work and
within knowledge." But they are restored specifically so as *not* to move
these categories over a line of reason. They aim, instead, to trouble that line
through an understanding of the entangled relationship between the social
and the natural (and all the other levels) that draws both on a feminist epistemology *and* on the daily experience of women scientists. I put this here as
a marker and reminder for the following discussion.

Let me close this section with one final extract—it comes from an interview with a postdoctoral neuroscientist who I described earlier as being
quite downbeat in her account of what neuroscience was or what it could be.
But what's interesting is that, immediately after that discussion, she began
to talk herself into a slightly more positive view of what brain imaging could
actually do—in particular, how looking at multiple levels was making for a
better neuroscience. She said:

Now, in brain-imaging methods and statistical methods, the way to look
at this information from brain imaging has evolved as well. So while now
one can infer, for example, when two areas talk to each other by looking at whether their activity is correlated. Again, it's hard to know which
one started. Because, for example, fMRI [functional magnetic resonance
imaging] is a very slow method, so it's still quite hard to tell what's the
time . . . the unfolding, in time, of activity. So you see two areas that seem
to be . . . whose activities seem to be correlated because they go up and
down together, but you don't know which one started. Um, of course you

can fill in some gaps with anatomical data—so now there are some ways to measure traffic to determine what is connected to what in the brain, and to measure the thickness of this tract. So you know how . . . you have an estimate of how fast information will go, for example, from one side to another. So this is called connectivity. So then you can put together these data from functional fMRI, where you see the areas activated with what you know from the connectivity in the areas, and then you may infer something.

What draws me to this extract, in particular, is that the researcher begins by describing limitations. But then gradually, and almost in spite of herself, she starts to give a really interesting, accretive account of how different components and levels ("anatomical data"/"the thickness of this tract"/"how fast information will go") can actually be strung across the statistical and anatomical gaps in brain-imaging data. Even if "you don't know which [brain area] started," you can use anatomical measures to "fill in some gaps," get a hold of some measures that focus on "traffic," then "put these together" with your fMRI data—and suddenly things start to look a bit more solid and coherent.

I am specifically working to understand this ability of modestly witnessing neuroscientists to move between anatomy and behavior, think about myelination and psychology, work with computer science and family life, connect the expression of a gene to the conduct of a conversation, and so on. The seemingly intractable incoherence of an autism—strung as it is across different epistemological, disciplinary and corporeal levels—might also be read as the ability of the new brain sciences (through the application, sometimes, of a relational, gendered, and modestly witnessed scientific labor) to work with, on, and through things that manifest across different scales. This begins to explain the *contingent stability* of autism. In using this term, I call attention to both the spectrum's implication in a particular sort of neuroscientific ordering and also the fact that to be implicated in such an ordering might sometimes be a condition of *being present* in the first place.

TRACE IT UP

I return to an interview first mentioned in the introduction. It comes from a conversation that I had with somebody working in developmental neuropsychiatry. In common with a lot of the people I spoke to, he was particularly interested in finding the brain basis of some of the core symptoms of autism. In the early part of the interview, he responded to my question about what had sparked his interest in this area by saying that autism was a developmental disorder with pervasive symptoms, and it seemed reasonable to guess that these symptoms might be amenable to his methods, which included both EEG and fMRI. Okay, I said, but given all that we knew about the variability and even the ineffability of autism (and so on), isn't the interesting thing about autism the degree to which it is strangely *unamenable* to these kinds of methods? He said:

> If the symptoms are manifest in terms of relationships with others, differences in perceptual functioning, differences in motor functioning, differences potentially in responses to social and concrete stimuli—then I have access to measures that measure those things. If I follow a line of thought which would say that whatever is going on in autism can be understood by a model of perturbed biological functioning, and if it seems at least halfway plausible that atypical biological functioning may be reflected in things like functional MRI, blood flow using SPECT [single-photon emission computed tomography—another brain-imaging method], or EEG recording, [or] Event Related Potential recording, then I have access to those. If, for instance, it seemed to me—because obviously, as you say, one has to have one's own personal history and pathway—if it had seemed to me likely that the pathology there is associated at either, on the one hand, a purely societal level or, on the other, at a purely genetic or molecular level—then I don't personally have access or expertise in those areas of research.

Here, much like some of the discussion I set out earlier in this chapter, he seems to place his (mostly brain-imaging) methods at a distinctive midpoint between the molecular (below) and the societal (above). His methods are those that can measure the functioning of organs *in between*—on the basis that there's good reason to believe that what's going on at the molecular and societal levels will be registered in some kind of "atypical biological functioning" in the individual brain.

For me, though, this raised the question of how a line of coherence was actually maintained *between* those levels—or how someone doing a brain-imaging study on high-functioning autistic adults knew they were dealing with the same thing as their colleagues diagnosing severely autistic children in a family clinic. Could a diagnostic entity really maintain a sense of coherence amid such unknowns? He said:

> That's a big question. Um, I guess the answer is, yes, I would think that something certainly could. And it could probably travel in both directions. To take the one that I think is less relevant to my own work first: if, for some reason, at a societal level, people treat you differently, so they treat everyone with red hair differently, then there's research now that demonstrates—say, you know, they abuse and ignore someone with red hair—then that does exist as brain-biology changes. There's research on, you know, different field altogether, but people who've been subjected to various kinds of abuse, from early childhood onward. So—it's not quite societal, but yeah, if you're a victimized group, then that can affect biology. [. . .] And that biology can be manifest in terms of behavior; it can be manifest in terms of hormonal function in the brain, and in terms of the functioning of particular regions within the brain, and there are studies that have shown that.

At stake, here, is a loop from some kind of phenotypic abnormality (for example, red hair) to society and the environment (different treatment of people with red hair), to anatomy and the body (brain changes on the basis of this treatment), and back out to behavior (a whole host of clinical symptoms that are produced by those brain changes).

I am trying to fix attention on the relation between this neuropsychiatrist's ability to draw, participate in, and—to use Haraway's term—figure such loops *and* the holding-together of autism as a somewhat stable diagnostic category across these different layers of knowledge and experience. It is noteworthy that this account is not only an acknowledgment or appreciation of "all the levels"; it also describes an experimental practice in which those levels are explicitly drawn and held together: here is neuroscience as strung-out and figured, looped and networked, collaborative and compromised—until enough strands of knowledge get laced around something that we can begin to recognize as *autism*. Later, he put it even more concretely:

Again, moving outside autism temporarily, there are clear isolated and probably rather rare, pure genetic lesions that have been defined in the laboratory that disturb language functioning. Then even if the rest of your brain, if you like, works fairly well, people behave differently to you. Your experience growing up is different because you have this molecular biological deficit that inhibits language development. And that will then start to interact—and it will lead to, or at least, you could easily develop this [*inaudible*] which would lead to biological differences in the way the brain develops, cognitive differences, interpersonal differences, and potentially societal effects. So that you can, without doubt, trace it up—now, not necessarily very easily. But you can.

I mentioned this extract in the introduction, but let me repeat it here nonetheless. This neuropsychiatrist's proposal is that even if you take the "purest" biological phenomenon you can think of—some genetic phenomenon that produces a distinct, predictable brain lesion, in an area unambiguously associated with some feature of language, and one that has very well-described effects—even *that* will inevitably change as the affected person goes out into the world and interacts with other people. And that change will be measurable at the level of the brain. So what you do, of course, from any one level, is you "trace it up." You move forward on the basis of that tracing.

He went on to describe, for example, attempts to reduce the "stigma" of mental disorder as a way of intervening in mental health, because you could trace those social effects to the appearance of the disorder in the brain and even potentially analyze the social effects using brain measures only. This image of the trace shows how this careful neuropsychiatrist works and moves between different levels, neither through guesswork nor construction, nor by simply following a natural path of discovery. He works rather through careful accretion and collaboration across different levels of knowledge, practice, and experience. The very same description was used by the PhD student, who I quoted earlier on the potential relationship between social priming and myelination. Later on in that conversation, she said:

I'm feeling like I'm contradicting myself a lot . . . um . . . um . . . because I was previously saying, "Oh, the brain's really important, I'm really interested in the brain," but I think that you need it all. You need behavioral experiments and you need to know what's going on in the brain, and you need to know, well, it's an advantage to know about genes, about stuff that's inherited, compared to stuff that you've learned. It depends what your question is though—like, what do you want to know? Like, do you want to know why individuals with autism don't exhibit this social modulation of imitation, if that were your question: Why? Then I do think that you need to trace it back through development, and you need to take into account inherited biological stuff and also experience, because [otherwise] you're not going to get at the question of why.

Again, I want to draw attention to not only the desire to "have it all" but the degree to which this *having* is associated with a careful, deliberate, and active *tracing* practice. This is a practice in which clear lines of connection are established between development and social modulation, on the one hand, and genetic inheritance and brain biology, on the other. For this researcher, to know the *why* of autism is to help trace the lines between these levels. It also means collaborating with the other agencies that are at work within them (like "development" and "inherited biological stuff"). Tracing becomes precisely the goal of the rational-responsible neuroscientist in search of some ambiguously constituted scientific object. And her sense that in describing this process she is somehow contradicting herself captures very well the way in which this kind of uncertain neuroscientific labor is not always intuitively obvious.

I am mindful of the affinities of my account with what Bruno Latour (1987: 236–37) has described as "immutable mobiles." These are transportable resources and agencies (in this case, we might say, journal articles, electrophysiology, red-haired kids) that are solid and well-regarded enough to maintain their shape. But they can also be arranged by resourceful "centers of calculation" into the kinds of networks—Haraway (1997: 267) might say, "sociotechnical orders"—that sometimes have stable scientific objects at their ends. Accounts that follow Latour's rubric often position themselves at quite a late point in the stability of these mobiles (among Latour's own cardinal examples are the maps of Ferdinand de Lesseps and the astronomy of Tyco Brahe), and this can sometimes create the impression of a self-conscious (even a bit self-satisfied) "actor-network theory" (ANT to its adherents)—in which everything is already achieved and (perhaps counterintuitively for those of us basically sympathetic to a *second empiricism* in the social sciences) in which strung-out, sociotechnical orders of achievement can look even a bit *too* stable.

More to the point: such a post hoc descriptive practice can begin to look like a science of the already obvious or a history of the winner. I am just not convinced that such a move is well suited to the shifting plates of the contemporary life sciences. "An object," in the manner of an actor-network theory account, at least as the actor-network theorist John Law (2002: 93) has put it, "remains an object while everything stays in place and the relations between it and its neighbouring entities hold steady . . . the job of ANT is to explore the strategies which generate—and are in turn generated by—its object-ness, the syntaxes or the discourses which hold it in place."

But the neurobiology of autism, like many objects of the contemporary life sciences, and not at all like the movement of the stars, is not yet so stable, nor are the generative strategies (if strategies they are) and agencies so well worked out. Things are not in place. Nor, for that matter, is it so obvious that emplacement is actually what's at stake. So while I draw on some of Latour's very useful descriptive apparatus for talking about scientific facts below, I also want to restate my commitment to Haraway's elaboration of facthood. It is the case for *my* data, as it is in Haraway's discussion, that scientific witnessing is not at all associated with finality or achievement (here, Haraway's impatience with the more militaristic rhetoric of some self-described ANT accounts is evident), but instead with forms of relay, exchange, multiplicity, complexity, and entanglement (Haraway 1997: 268). I draw on the image of Latour's "immutable mobiles" in what follows, but my emphasis is on a still-in-process tracing and relaying, which requires a bit more uncertainty than such accounts sometimes admit.

Here is another description of the process. The contributor doesn't talk about tracing, but he uses a related metaphor of "shuffling things together." What is powerful about this account is that the researcher begins talking about one—then popular—model of autism but quickly begins to recruit other agencies into a potentially stable account of just how such a model might be convincingly traced together, beginning with his own speciality (MEG brain imaging):

So, for instance, one model [of autism] is that there is inherent underconnectivity. So autism is seen, according to this model, as something where locally brain areas interact, but they're lacking the big picture if you like. So they're lacking the ability to connect over large distances. You can then take the MEG result and see how this fits into the model. So it is, if you like, it is a bit of an iterative process. You have correlations first from other task, from other experiments, perhaps even from other populations of subjects you have theoretical models, and you can now start shuffling things together [. . .] Take this example—say the [cognitive] theory is holistic, local processing in autism. You could be inclined to attempt to match that one to [another] one by saying, "Well, holistic refers to global synchronicity, and local or piecemeal processing refers to local processing." But we already know we can't be that simple. But the idea is, here, again, is to make as far as possible some prediction, even if they are conceptual in nature, from the theoretical models, and see how this can or cannot match.

This interviewee was a physicist by training, and he spent some time impressing upon me his ideas about the relationship between EEG and "thought," and his admiration for the work of philosopher of mind John Searle. Here we can see how the "hard" neuroscientist thinks about making different cognitive models speak to his data. He tests potential orders for Latour's categories of *mutability* ("Well, holistic refers to global synchronicity"), *stability* ("see how this can or cannot match"), and *combinability* ("you can now start shuffling things together").

It is not at all clear how (or if) such an account would finally hold together, or who or what would need to be involved. But this early, speculative discussion gives us, I think, a strong enough sense of how it might happen. It lets us see how an autism of all the levels of understanding might be more closely drawn together, and it gives us a sense of how shifts might occur in the contingent stability of such an object. Another senior scientist said:

What I compare [autism] to, what I compare it to is like a tapestry like, there's, uh, *La Dame à la Licorne* [*The Lady and the Unicorn*] at the Musée de Cluny in Paris—a very colorful tapestry, a medieval tapestry. And you cover it with black plastic, and what you do is you punch some holes in it, and shine a light through, and sometimes you'll see blue, and sometimes you'll see gold, and sometimes you'll see a little bit of pattern. And all scientists are all kind of saying, "No, autism is blue, no autism is red, no autism . . . ," you know. And we haven't punched enough holes even to see a decent bit of it—even to say, "Well, actually, there's a unicorn."

The unicorn—an image, of course, that already brings into play the mutability of the mythic and the real—is presented here as a kind of tentative agency, at the center of a range of practices of making visible. But it is only something that becomes meaningful, and identifiably there, when all of the different elements—blue, gold, "a little bit of pattern"—are carefully strung together. The point is that the unicorn is not only something waiting to be discovered (if only these researchers could somehow find a method to tear off the black plastic!). The unicorn is instead both made up of, and revealed by, the combination of ongoing accounts of blue, red, and gold. The tracing together of all these different elements precisely draws into question the helpfulness of distinguishing between the work of tracing these elements and the unicorn itself—the trace behind the veil. The claim that, one day, we might realize that "there's a unicorn" is not a vision of final insight: it is the tentative hope of an accretive and combinatory, tracing neuroscience.

A tapestried unicorn is as good an image as any on which to end this chapter. Entwined within it are the tangled tropes of artifice and discovery that have run through this account as well as images and memories of the kinds of uncertain, following-and-marking labor that I have been trying to describe for the neuroscience of autism throughout the book. As an ambiguous, veiled amalgamation of the really-real and the not-quite-real, joining together the work of revealing and the labor of making, tracing the unicorn, as this interviewee makes clear, is no small thing. But it is an image that we might draw upon and think with, finally, to see not only a neuroscience that is attentive to "all the different levels" and also not only an autism that is strangely and problematically dotted across different ways of thinking, knowing, and experiencing. This image helps us to think about what is really at stake in the contingent stability of autism—namely, an ongoing, processual, tracing practice: one that is inherent to the relational and combinatory logic of the new brain sciences, which works by slowly recruiting different agencies and drawing them together, and which carefully labors, in the end, to make some kind of coherent order out of them. This graduated, ambiguous, and modest image, which bespeaks not only care but achievement too—this is precisely how I think about the strange multiplicity of the neurobiological account of autism.

POSITIVITY

I began this chapter with a desire to concretize some of my more general commitments to the neuroscience of autism, which have run throughout this book. These commitments came from an insistence that it should be possible to think more or less sociologically about the tensions, ruptures, and ambivalences that run through neuroscientific research on autism, without using these differences to undermine an emerging neurological account of how autism is either *of* or *in* the brain. In this chapter I have tried to show how, even if autism is a disorder strangely adrift between very different sets of scientific practices and assumptions (and even if it is not always clear that neuroscientists can indeed move as easily between these levels of discussion and practice as they might like), still we might begin to see their quick and shifting creativity, and the sense of solidity that it produces, as reasons to think more carefully about the specific elements of both the multiplicity and the movement in question.

I think these things are worth knowing. As I alluded to in the introduction, there has still been too little—I draw on Eve Kosofsky Sedgwick's (2003) term—nonparanoid discussion in the sociology of the brain sciences (and indeed in studies of science and technology more generally) vis-à-vis the careful ways that novel diagnostic entities are held together by scientific practices and epistemologies. I have been trying to show, to put it in the most simple terms possible, that the discovery of some unexpectedly human, complicated, and contradictory elements within an experimental practice is actually a sign of nuance and care—not (as our colleagues sometimes still have it) of epistemological naivety or of the ineluctable force of social context. The fundamentally reparative gesture of thinking otherwise does *not* require us to become vacuous cheerleaders for the new brain sciences. But it does mean committing to a model of a criticism that basically wants the object of its own discussion to do well. The philosopher Graham Harman has reminded us:

> An old maxim states that there are two kinds of critics: those who want us
> to succeed, and those who want us to fail. Debate is always tedious when
> conducted with persons of the latter kind. Wherever we turn, they are
> popping balloons and spilling oil on the floor; we find ourselves confronted

not only with arguments, but with unmistakeable aggressions of voice and physical posture. Yet such gestures of supremacy yield no treasures even for the victors, and somehow always seem to solidify the *status quo*. It is analogous to "critiquing" long distance buses by puncturing their tyres, assuring that no one leaves town and nothing is risked. (Harman 2009: 119)

This is perhaps a bit more polarized than I would like, but it still stands, fairly accurately, for one of the central ambitions of this chapter, which has to shift debate away from tedious fault-finding even while looking through structures of difference and ambiguity.

Again, I am aware of the difficult politics that I am stepping through. If, as I conceded in chapter 3, there are indeed reasons to worry about some of the tendencies emerging in the new brain sciences (although this may owe more to a popular than a scholarly literature; see Johnson and Littlefield 2011), these questions become even more acute in the face of a multiply-contested diagnostic category like autism. As noted at the end of chapter 4, autism remains mired in important contests about the relationships between neurology and identity, the passage between difference and legitimacy, and the definition of what gets to count as either *social* or *neurological* in the first place. The politics of these questions are deeply intertwined with the kinds of laboratory-working that I have described. To talk about the neuroscience of autism in terms of a simple-looking reparation, in the midst of such arguments, risks looking glibly avoidant of the hard social and political questions that impinge on scientific practice.

Part of what I have aimed to show, however, is precisely the inseparability of sociopolitical and neuroscientific gestures in autism research. I have tried to show how describing one is always also to begin a conversation about the other. I hope to have shown, too, the kind of neuropolitics to which I am indeed committed. This is embedded in a commitment to the ongoing making-up of a neuroscientific research practice, and also to a set of practitioners, without always trying to settle debates, to reduce levels, or to announce victories; it is a practice, on the contrary, that builds complex accounts of the human world, and puts together interventions for that world, through the use of carefully wrought, and deeply cosmopolitan, alliances, coalitions, and affiliations. It does so on the basis of a scientific experimentation that is as corporeal, affective, and ecumenical, as it is careful, methodical, and tactical.

CONCLUSION

The Pursuit and the Mark

WHERE DOES THIS LEAVE US? THE TITLE OF THIS BOOK IS *TRACING Autism*. But I am aware that my attention has tended toward the first half of that phrase, *tracing*—and thus toward the talk of scientists engaged in the labor that I have tried to describe. In so doing, the book has edged away from the latter half of the title—away from the set of experiences, definitions, dispositions, and affects suspended in that noun, *autism*. This was not a conscious decision: when I set out on this project, I thought there would be a strong autistic voice in the story that I was going to tell. Certainly there is too often a willingness, within social and (especially) philosophical analysis of psychological, psychiatric, and neurodevelopmental categories, to mobilize diagnoses and cases too lightly—as if such cases did not in fact band themselves awkwardly around rich, varied, and contested assemblages of actual human experience.

This is particularly an issue in discussions around autism, where the intersection of scientific knowledge and autistic life has a long history, and not always a happy one. Moreover, the hubbub of contestation that has grown up around these intersections does not only relate to an unfortunate psychoanalytic past: it extends well into the present, is audible in contemporary genetic and neurobiological research, and winds itself through the voices and presences of many different autistic people as well as those who identify as family members and allies, not to mention the global field of clinical and research managers, a growing array of NGOs and advocacy organizations, and the wider set of research funders, policy makers, and so on. Across these very different political and cultural contexts, an important debate is under way about how autism can or might be talked about—about what constitutes the proper field of operations for those who would do research around it. I am aware that this book will do little to reassure those who (rightly) hold the view that there is far, far too little sense of an autistic presence, or an autistic voice, in research contexts that presume to speak of and to *autism*.

The truth is that quite early on in my interviews with neuroscientists, it became obvious that the accounts that circulated across them—how these scientists were mobilizing, thinking about, and figuring autism within their own practice—were already deeply contested, entangled, multifaceted, contradictory, and so on. It seemed to me that there was a particular story about neuroscience here, one that was not yet told but that would be well positioned to draw out a deep-set sense of ambiguity and contest within neurobiological research. Autism was and is profoundly implicated in this story. (Which is to say, and to answer a question that I am repeatedly asked but cannot empirically demonstrate: No, I don't think I would have gotten the same material from talking to neuroscientists concerned with bipolar disorder, depression, or attention deficit hyperactivity disorder—or one of the many other, no-less-vexed, developmental, psychological, and/ or psychiatric categories through which neuroscientific knowledge and practice are getting made and remade today.)

I hope, very sincerely, that a deep commitment to thinking seriously and carefully about autism, even via second-order accounts from scientists, comes through in this book. But it remains the case that this is a book more about neuroscience than it is about autism. I did, for a while, nurture a hope of extending this work into a larger engagement with autistic communities and the clinical practices that seek to intervene in them—but that fell victim to the caprices of funding, time, career, and so on (the unhappy details of which I can probably talk about more in person). Let me begin this short conclusion with an attempt to be a bit more declarative about autism—as both a category and an experience—and about its relationship to neuroscience in particular. After that, I want to say something more direct about the broader literatures that I have been able to get at (I think) by thinking with autism: first, neuroscience and psychology more generally, then also the sociology and anthropology *of* those fields.

NEUROSCIENCE AND AUTISM

I return, very briefly, to where we came in: the search for an autism bio-marker. At the end of a book that has staked its contribution on the gen-erative value of ambiguity, and also being an obvious nonexpert, I am a bit reluctant to be super declarative about this. Still, it seems to me that, if anything has been learned from these conversations, it is that a single-minded quest for a definitive, brain-based biomarker of autism—the sense that such a marker is likely to arrive, and indeed should it arrive, that it will have a great deal to do with the lives of actually existing autistic people—might be misguided. Whatever we know about autism (neurobiologically or otherwise), we know that it is expansive, capacious, and varied. This might not only be because we have not yet properly understood its contours but because it comes loaded, in advance, with such developmental and cogni-tive variability, that it already exceeds the methods and concepts we have painstakingly developed for understanding our collective idiosyncrasies and differences. Already I can see the eye rolling from my neuroscientific interlocutors—*of course* no one is looking for a single, brain-based biomarker. Well, maybe not. Still it seems to me that many large-scale autism-research initiatives come freighted—even tacitly, perhaps without anyone ever actu-ally saying as much—with precisely such a hope.

This leads to the second thing I want to say about autism: something may be biologically real enough and yet impossible to pin down biologically. In other words, and being fairly simplistic about it, to be reasonably certain—and to act on the basis of that certainty—that there is an organic condi-tion called autism does not in fact presuppose the ability to isolate a causal genetic or neurological function that always and inevitably signifies *autism*. If this is obvious to some, well, it wasn't—and still isn't—to me. And none of that means that people won't still *try*—especially via more sophisticated ways of thinking about biological markers. One possible future (perhaps the most likely future) for the neuroscience of autism is that rather than trying to simultaneously describe everything that we today bind within that word, a clinical biomarker derived from neuroscience will enact its own cut when we have a strong enough account of some meaningful collection of neuro-biological or genetic factors, which allow us to describe enough of the vari-ability to do something useful for enough people.

Rather than trying to understand, at the neurobiological level, everything we currently describe as autism, we might eventually come to understand *autism*, neurobiologically, as that which we are able to usefully describe. Certainly there are virtues to such an approach. Yet it seems to me, as a nonscientist outsider, that what is at risk of getting passed over here is the tendency, within the neuroscientific disciplines, to *extend* rather than to cut—to pile more descriptions on top of one another, rather than trying to isolate the one that *really counts*. Do we lose anything when we take the fundamental task of imaging technology, vis-à-vis problems in neurodevelopment, to be a finer, more parsimonious diagnostic apparatus, rather than the inculcation of a thicker, more substantial descriptive genre? One way (and I guess the way favored by most autism researchers) of dealing with the inherently capacious quality of neuroscientific practice is to think about a more expansive universe of autisms, subautisms, different genetic and behavioral syndromes that today we *just call* autism, and so on. But what if rather than focusing on the enactment of these distinctions, we took the expansiveness of neuroscience as an impetus for understanding autism (and maybe not only autism) as an undoubtedly biological and singular phenomenon, but one that is nonetheless irretrievably suspended within a series of different experimental results, findings, stories, markers, experiences, articulations, feelings, and so on? I say this as someone who doesn't have to do that research (or seek funding for it), but what if the point of neurobiological research wasn't circumventing or resolving this conceptual tangle but learning to live with it?

The third thing I draw into question is the idea that the brain is necessarily the best place to start thinking about intervening in autism anyway. As the psychologist Liz Pellicano and her colleagues (Pellicano, Ne'eman, and Stears 2011) have pointed out, funding already disproportionately favors etiological research rather than things that might benefit autistic people in the here and now. This is not at all to be dismissive of the careful and useful work that neuroscientific investigators do. Nor is it to instigate a zero-sum discussion of what best to do with available funding (a world that robustly supports neuroscientific research as well as investigations into more mundane interventions aimed at improving day-to-day life for autistic people should not be unimaginable). It is simply to wonder at the weight of cultural expectation that we burden neuroscientific investigators with. It seems fairly obvious to me (I am far from the first to say it) that there is a range of more everyday interventions that would more likely benefit autistic people in the shorter term—from autism-friendly educational and social spaces, to wider public awareness of how diverse autistic experiences might be better understood, to the provision of high-quality adult social care (and transitions to that care) for autistic people who need it, and so on.

I am straining to avoid a crude dichotomization of high-tech bioscience (which is indeed expensive as well as sometimes necessarily and appropriately speculative) and on-the-ground interventions (which are rather more likely to have a real effect in the short term, albeit they are less likely to excite prize committees, university press offices, *Nature* reviewers, science journalists, and so on). Nonetheless, from a research strategy point of view, there is perhaps room to be more circumspect about the role that neurobiological (and genetic) research is likely to play in improving the day-to-day lives of autistic people—at least in the shorter term.

Indeed, to briefly return to a long-standing hobby horse of my mine (Fitzgerald and Callard 2014; Callard and Fitzgerald 2015), we need to see a great deal more interdisciplinary collaboration within and around neuroscientific work. I don't just mean collaborations between psychologists, neurobiologists, cell biologists, and so on (this happens frequently enough). I mean collaborations with historians of science, with sociologists and anthropologists, with ethicists, with literary scholars, and many others who would actually have a lot to contribute to the generative potential of the perspectives I have described (setting aside the question of research collaboration—and I mean *interdisciplinary collaboration*, not *engagement*—with autistic people themselves, which is a different but equally vital issue). Might there not be some taxonomic gain, for example, for the kinds of debates I described in chapter 1, from some investment in the history of psychology? Could working with ethicists or literary theorists open paths for neuroscientists to productively channel the affective energies that I descried in chapter 3? Could a sociological or anthropological perspective not help to invigorate, or operationalize, the concerns about brain imaging that I described among neuroscientists in chapter 2?

Perhaps this seems fanciful. But I am reminded of the work of the wide-ranging French sociologist Brigitte Chamak, who, with her colleagues (Chamak et al. 2008), uses anthropological methods to contrast autistic people's representations of autism with scientific accounts (showing that, in autistic people's accounts, sensory idiosyncrasies and perceptual differences take priority over the autism triad that for a long time defined the condition among clinicians). I am also reminded of the work of such researchers as Morton Ann Gernsbacher, Michelle Dawson, and Laurence Mottron (2006) in Montréal, who have pushed against the psychological and psychiatric commonplace that autism should be thought of as something "harmful." Extending perspectives from the neurodiversity literature into the psychological laboratory, scholars such as these show how autistic intelligence is underestimated by normal psychological tools (Dawson et al. 2007).[1]

I am also reminded of work of Liz Pellicano and her colleagues in London, who, among other interventions, have used qualitative methods to understand autistic people's views on how to actually describe autism (Kenny et al. 2015), while working collaboratively with autistic adults to move "beyond the boundaries of a neurotypical culture" in the context of design research (Gaudion et al. 2015: 49). This is hardly an exhaustive list. And of course there are important differences between these papers—I don't straightforwardly describe them all as either interdisciplinary or collaborative. But what holds them together, I think, is a willingness to push against the disciplinary confines of *normal* psychological work on autism, and to do so in a broadly collaborative way, but nonetheless to work pretty squarely *within* the experimental, fact-producing traditions of contemporary (neuro)psychology. They begin to show us that we could think much more seriously about how methods from anthropology, and history, as well as the neurodiversity and self-advocacy literatures, can do something other than undermine psychological and neurobiological assumptions (even though such undermining is often important work too).

NEUROPSYCHOLOGICAL NATURECULTURES

On the subject of collaboration, let me say something concrete about the sociology and anthropology of the new brain sciences. Basically all I want to do is lament the absence of significant conceptual development in these subfields in recent years. Despite significant shifts in neuroscientific thought styles, social science anxiety, despite some notable exceptions (I think especially of Rose and Abi-Rached 2013), continues to oscillate around the supposedly reduction-ist intention of neuropsychological and neuropsychiatric research. This is a worry, to be simplistic about it, that some complex, differentiated, and entangled aspects of the human social world are at risk of being reduced to the bare life of the brain (see Martin 2000, 2004; Ortega and Vidal 2007; Choudhury, Nagel, and Slaby 2009; Gillies, Edwards, and Horsley 2016; De Vos and Pluth 2016). I am not dismissive of such concerns nor of the scholarship underpinning them. Let me go further and say that ten to twenty years ago, suspicious and paranoid reading of the neuroscientific literature was probably very much warranted. If you focus only on the public pronouncements of major neuro-scientific figures, or press releases that accompany articles, or books written by scientists for a broad public, you will likely think them still justified *today*. I remain alive to the possibility that, another decade or so down the line, my own currently hopeful view will be shown up as a shallow and naïve enthusi-asm—a description that, no doubt, some would already append to it. (Indeed, my optimism, such as it is, remains a job of work and often a trying one; it comes neither easily nor naturally to my reading of the neurobiological scene.)

But there is no grand claim about the neuroscientific future itching to get out of this book. My entirely minor ambition—this is not a false modesty—has been to show that when you talk to neuroscientists who are engaged in the *cerebralization* or *neurobiologization* of some complex psychological or psychiatric diagnosis (such as autism), you are unlikely to encounter a con-fident language of monolithic neuroreduction. What you will probably hear instead is a much more complex, ambiguous, and uncertain discourse. You will encounter a way of talking about neuroscience that is characterized by an ability to live with, and work through, forms of entanglement and complexity—a way of talking that can't really be described as a desire to reduce (let alone a method for reducing) one kind of thing to another. And this is important. Not least, it means we have to significantly recast what we think "neuroscience" is, as (if I might use such a term) a social fact. But it also requires more significant theoretical and methodological recalibration, at least for those of us interested in the social life of the mind and brain. This is a recalibration in which neuroscience is not simply an aspect of the social world to be analyzed (good, bad, indifferent). Rather, it is one in which neu-roscience becomes in fact a valued coproducer of our knowledge *about* that world, a collaborator no less nuanced than we sociologists and anthropologists ourselves, in our ongoing and shared attempts to understand the kinds of collective experience within which our social, cultural, affective, psychological, and cerebral lives take shape.

As I said in the introduction, *I went looking for the monolith*. But when I asked a group of autism neuroscientists what autism *is*, they said variously that it was a "biological truth" but also a diagnostic "umbrella of convenience"—and anyway, there was probably something out there called *autism*, at least insofar as they could feel it when they talked to someone who had it, and they knew it when they saw it. When I asked them to talk about their relationship to the basic practices of neuroscience, they sometimes said that neuroscience was indeed a powerful framework for psychological and psychiatric research, and that neuroimaging in particular gave psychologists and psychiatrists access to the organ that interested them. But they also said that neuroscientific measures were really partial and simplistic, that neuroimaging was actually quite biologically disappointing—that it always gave positive results no matter what you did, and that its growth might have as much to do with careerist self-interest as it did intellectual discovery. When I write in this book about the ongoing presence of ambiguity and uncertainty at the heart of some neuroscientists' accounts of their own practice, when I call for more sociological attention to—and positive engagement in—the neuroscience that prevails at this level, this is the kind of thing I mean.

I used the word *trace* to understand what was going on here. What I liked about this word is that it seemed to hold together two contradictory sets of meanings. On the one hand, there was the sense of tracing as active, generative work: *to trace*, in this sense, as a scientist, is to take one's course or make one's way; to dance; to travel or range over; to make out; to pursue; to "plait, twine, [and] interweave"; to "braid" (Oxford English Dictionary 2016c). All of this will be familiar to readers of literature within the social study of science, even if it remains a surprising term (to me, at least) for a midcareer neuropsychiatrist to fall back on. But *trace* has another set of meanings, too, which seem to already subvert some of the relations implied in the first set. Here, a *trace* might also be a path or a line; a "series of steps in dancing"; a mark or an impression; it can refer to footsteps; to a "track . . . beaten by feet"; even to "vestiges or marks remaining and indicating the former presence, existence, or action of something" (Oxford English Dictionary 2016b). What are we to make of a term that elides, with apparent ease, the difference between the travel and the path, the dance and the feet, the pursuit and the mark? What work might such a word start to do for us, if we really are going to get over some of the ontological and epistemological simplicities through which many still, today, talk about the neurosciences?

I want to avoid overinterpretation. I have talked about *tracing autism* to get some conceptual purchase on an important unifying element within my interviews—namely, the way in which neuroscientists' talk did not usually concede a sharp divide between (1) the sense of ambiguous entanglement that they were working through and (2) the potential, all the same, for a singular account of the neurobiological substrate of autism. *Tracing Autism* is not an attempt to describe a coherent system. Nor does it offer any kind of theory—at least not an original one—of how neuroscience works. In chapter 2, I affiliated my focus on the trace with Karen Barad's (2007: 33, 175) "agential realism," which has already recognized the "intra-active" priority of entanglement while creating space for the independent existence of "agentially cut" nonhuman things all the same. In chapter 3, I deferred to A. N. Whitehead's (1979: 49) account of subject-object relations, which not only troubles the distinction between being subjectively apprehended and being independently existing but even makes the two conditional upon one another. In chapter 5, I proposed a relationship between the act of tracing and the generation of what Bruno Latour (1987: 236–37) has called "immutable mobiles," which is his name for an identity between carefully strung-together networks of things, and things themselves.

As this somewhat parasitic schema shows, my use of tracing is less metaphysically ambitious than any of the works upon which it leans. It is, at heart, a way of talking about some similar intuitions, and practices based on those intuitions, drawn from different parts of my interviews with neuroscientists. It is important to me that the conceptual labor comes out of the interviews themselves: what I want to call a *tracing neuroscience* not only describes a rich and complex method for thinking with and about entangled worldly agencies; it conveys a deeply sophisticated *internal* apprehension of, and affective intersection with, the awkward, hybridized, material-semiotic entanglements of contemporary naturecultures, that still go on under the signs of psychiatry, psychology, and neuroscience.

At the heart of a *tracing neuroscience* is a proposal to work—and to work empirically at that—through the entanglements of method, ideas, materials, and deductions that might be called *social* (or *cultural*), and the initiatives, feelings, processes, and conclusions that would more commonly be described as *natural* (or *scientific*). I draw, of course, on Donna Haraway's (1997: 56) concepts and terminology—especially her use of the word *naturecultures* to designate moments that "greatly increase the density of all kinds of . . . traffic on the bridge between what counts as nature and culture." For Haraway (2007: 25, emphases in the original), the fact of this crossing directs attention to the proliferation of spaces "in which all the actors become who they are *in the dance of relating*, not from scratch, not ex nihilo, but full of the patterns of their sometimes-joined, sometimes-separate heritages both before and lateral to *this* encounter."

What I have been trying to gesture at in this book is a kind of neuropsychological *natureculture*, in which scientific work proceeds almost explicitly through such forms of relating. In chapter 1, for example, I showed how autism neuroscientists are sensitive to the scientific and political differences between thinking of autism as a natural fact and thinking of it as a diagnostic category that serves a medical bureaucracy. What was interesting is that my interviewees did *not* sharply separate these natural and cultural categories, preferring instead to mingle the two. I showed how these neuroscientists indexed the definition of "what autism is" through categories of *enigma* and *sensation*—creating a bridge between the natural and cultural definitions of autism by figuring autism itself as a kind of mystery, and concrete knowledge of autism as a sensitivity *to* such mystery. In chapter 2, I showed how they did not enact a clear separation between the scientific promise of natural facts generated by brain scanning and the cultural presence of a collective anxiety about this practice. What was interesting, in this case, was that the latter concern did not function as a normative device to separate good neuroscience from bad. Instead, the generation and sustenance of a neuroscience that held onto senses of *both* promise and disappointment maintained a traffic between the scientific certainty of the former and the cultural *outside* indexed by the latter.

It is certainly uncontroversial now among researchers to say that autism is neither an exclusively social or natural category—but one that in any event is very widely dispersed genetically and neurologically and also that, in its current description, has inevitably had some of its natural edges brushed off for clinical convenience. My claim for a *neuropsychological natureculture* is a bit tangential to this. I argue that a *tracing neuroscience* of autism does not always make a great distinction between natural and cultural facts. Indeed, in its attention to complex developmental diagnoses, I argue that a tracing neuroscience is quite directly *productive* of what cultural theorist Lisa Blackman (2005: 186), in her analysis of the psychiatric and psychological sciences, has described as a "dialectical interchange" between natural and cultural forms. The historical, cultural, and biological complexity of the autism spectrum is then no limit to neuropsychological research. Indeed, it is precisely via the capacity to (affectively and otherwise) initiate and apprehend exchanges across these registers—cultural, biological, and so on—that scientific work *actually happens*.

INTERIORITIES

Let me say something, finally, about the contribution of this book to the disciplines to which (most of the time) I find myself to be subscribed—sociology, on the one hand, and science and technology studies (STS), on the other. Throughout this book I have tried to enact a particular kind of conversation with the neurosciences. By conversation I mean a practice of reading, listening, thinking, and talking *with*, which holds suspicion in abeyance, and which is therefore quite uninterested in going behind the scenes, in putting everything back into its contextual box, in getting a hold of what's *really* going on—or whatever. The task is then not to draw forth the hidden ideological and material commitments that cohere within different kinds of agency, but to start with the assumption that spectra of ideological, affective, scientific, cultural, political, methodological, and material interest already exist in relations of *interiority* to one another. So the work is not about stratagems for getting behind the appearance of unity; rather, it is about figuring out ways in which different agencies have come to inhabit one another at particular moments, and of beginning to wonder what kinds of things are at stake within specific forms and moments of inhabitation.

I am not claiming any novelty for such a view. Indeed, it leans heavily on two emergent themes that are crossing sociology and STS as I write, which already hold together two ways of imaging different futures for these fields. The first of these is a developing interest, within a range of qualitative(ish) social sciences, in what is sometimes called the *biosocial*. This describes a growing commitment, among some social scientists, to not separating biological and social agencies from one another, either onto-logically (in that the biological and the social are not simply plugged into one another in organisms, as if they were formed of neatly separable domains) or epistemologically (insofar as pretty much all elements of organic life need to be understood as concatenations of biological and social propen-sity, rather than being divided into instances of one [genetics] or the other [power]).

This position has a range of sources, not least a view that the "postgenomic" moment in the biological sciences has produced a more committed atten-tion to the *emergent* properties of, for example, genetic and neurobiological predispositions, vis-à-vis the environments (physical and social) in which those dispositions take shape. The unhappy portmanteau *biosocial*, of course, rather unsatisfactorily covers a range of quite different positions. These range from the fairly traditional syncretism of a biological anthropology; to the sometimes deadening correlations of social epidemiology; to attempts (I will not name names) to follow the golden coattails of epigenetics, social neuroscience, or similar; to serious historical, philosophical, and sociologi-cal theorizations of what might actually be entailed, empirically, in refusing to disentangle the social from the biological.

Perhaps the most compelling account in this last category has come through Nikolas Rose's (2013: 4) analysis of a new formation in the human sciences—a development in which "no longer are social theories thought progressive by virtue of their distance from the biological." As Rose has pointed out, what announces itself today as the biosocial has a longer and stranger history than many would like to acknowledge, being rooted in the conjoined birth (and awkward separation) of biology and sociology in the early twentieth century (see also Renwick 2012). Beyond the announce-ment of epochs, Rose (2013: 16) directs our attention to a more modest empirical task—to the work of "seeking opportunities for a more positive relation to these new understandings of what it is to be human" emerging in fields such as social neuroscience and genomics. To do this, he impels us away from additional conceptual speculation of our own and, in the spirit of Georges Canguilhem's attention to the biological philosophies of life that characterized his own era, urges "empirical investigations of the operative philosophy of the biologists themselves" (ibid.: 22).

With Nikolas Rose, I have tried to position neuroscientists not only as desirable collaborators for the biocurious, but as authors, methodologists, and theoreticians in their own right. In so doing, I leave behind the vision of an imperialist hard-science-of-everything that sometimes haunts my colleagues. In its place, I am working to install a notion of neuroscientific concepts, practices, and methods that are much more capacious than we have been led to believe (not least by some neuroscientists themselves). There is something worth thinking *with*, rather than only *about*, here. Not only am I interested in offering a different idea of what neuroscience *is*; what I have been in conversation with, in fact, is a set of knowledge practices that are strikingly open to collaborative forays from qualitative and humanistic colleagues. At least some part of my interest is in what we might then *do. Together.*

For a lot of people in the neurosciences, it's not so easy—and maybe not so desirable—to cut different kinds of historical, political, biological, and social factors from one another. What would be possible if more in the social sciences were of a similar disposition? To that extent, this book can be read as some kind of commitment to the *biosocial*. But I am quite profoundly uninterested in laying claim to, or intervening in disputes in, the territory that surrounds that word. What I *am* interested in—and what I think this book in part contributes to—are closeup accounts of some quite specific practices in the life sciences; a fine-grained attention to what is actually methodologically and conceptually at stake for people working *in* those areas; and some kind of empirical commitment to the hard labor of parsing where exactly, and to what end, social scientific analysis might start to intervene in those stakes.

The other complex that this book is likely to get read through is the set of texts and attentions that have, in various ways, sometimes for quite different reasons, lately set themselves against *critique*—especially insofar as a self-consciously critical practice manifests as an ever-present suspicion of the natural sciences. Canonical here, in the STS world at least, is Bruno Latour's essay from 2004, in which he—alarmed by how skilfully critical arguments from STS were being wielded by climate change sceptics—pointed out that it is an excess of *suspicion*, not of confidence, that represents the greatest danger today. What is at stake for Latour, in overturning such distrust, is an anticritical attitude that produces a particular sort of realism—in his case, rooted in an idiosyncratic metaphysics of *concern*. This isn't really what I'm after here. My ambivalence about the critical attitude has a slightly different genealogy—coming through, on the one hand, the branch of feminist theory that moves through (and, in turn, reframes and contests) Eve Kosofsky Sedgwick's (2003) potent distinction between paranoid and reparative reading; and, on the other hand, through feminist science studies, not least the branch that has long been indebted to Donna Haraway's (1985, 1997, 2016) rich and ambiguous figurations of the contemporary life sciences.[2]

Recently the literary theorist Rita Felski (2015: 11) has wondered what kind of reading gets lost in the freighted desire for critical interpretation, which insists on placing a work of art squarely within a box of *context*—thus condemning it for its sociability and, in the same move, losing all that is "incandescent, extraordinary, sublime, [and] utterly special" about it. Aside from anything else, Felski (ibid.: 17) has pointed out, critique in this mode is "a poor guide to the thickness and richness of our aesthetic attachments." Yet such a wearying cynicism is, she writes, a good deal more mundane than we would imagine from the charismatic authority that attaches to its most skilled practitioners (ibid.: 47). *Nature*, moreover, is often the most put-upon recipient of the critical mood. "The most urgent task of critique," Felski (ibid.: 70) argues, "is to 'de-naturalize'—to turn what appears to be nature back into culture, to insist that what looks like an essential part of the self or the world could always be otherwise. It would be hard to overstate the pervasiveness of this antinaturalist rhetoric in contemporary scholarship."

Perhaps the real tragedy of critique is this: it genuinely misunderstands stances that are in fact utterly conventional as somehow cutting against the grain of contemporary thought. Who does not now weep to read, in any recent text, solemn declarations that a technology disciplines a body, that a science creates a subject, that a machine has a politics, an object a territory, a laboratory an ideology, a field a history, and so on, and on, and on, ad infinitum? Against such banalities, I want this book to join in the encouragement of a postcritical mood. What I have tried to show here is that interest and charisma might lie elsewhere—that there is something strange, risky, and unexpected, yet to be discovered, in the seemingly mundane work of tracing something else.

NOTES

INTRODUCTION

1   To take just a few prominent examples that occurred as I was working on this project: In 2010, Timothy Roberts and his colleagues (2010) at the Children's Hospital of Philadelphia proposed a brain-imaging biomarker for autism, using magnetoencephalography (MEG), to distinguish autistic participants from controls on the basis of a latency in processing sound. Finally, claimed the science correspondent of the *Daily Telegraph*, "researchers believe they have discovered a potential way of spotting the disorder in early infancy by scanning the brainwaves" (Alleyne 2010). In the same year, Christine Ecker and her colleagues (2010) at the Institute of Psychiatry, London, published a paper describing the use of magnetic resonance imaging (MRI), which they attached to a complex system of classification, to distinguish people with autism from controls across five morphological dimensions—an outcome that got its lead author invited onto the *BBC Breakfast* television program. The following year, Michael Spencer (2011), a psychiatrist at the Autism Research Centre, Cambridge, led a team that used functional magnetic resonance imaging (fMRI) to measure responses to seeing others' emotional expressions—a measure that, he hoped, might form the basis of a brain-based biomarker for autism. *The Daily Mail* was moved to describe this as "ground-breaking research" that "could pave the way for treatments or even a cure for sufferers of autism and Asperger's syndrome" (Daily Mail Reporter 2011).

2   Because it's not central to what I want to argue, I won't spend a huge amount of time distinguishing the different brain-imaging methods and what they do. But for many of the people I interviewed, functional magnetic resonance imaging (fMRI) was their main method. The specific claim of fMRI is to image the brain while it's actually *doing something*—using a measure of blood oxygenation as an index of brain activity. Hence the iconic status of fMRI within cognitive psychology particularly, where it can scan the brain and produce neural correlations while a participant performs a cognitive task such as looking at faces (see Beaulieu 2002 for a detailed ethnographic account). Other popular methods include magnetic resonance imaging (MRI), a form of medical imaging also used for structural scanning of the brain based (like its functional equivalent) on the relationship between electromagnetic fields and the alignment of atomic nuclei in the body (see Joyce 2008 for an ethnographic account); electroen-

cephalography (EEG), one of the more long-established forms of brain imaging, which measures electrical currents in the brain (see Borck 2008 for a history); and magnetoencephalography (MEG), which is related to EEG but based on measuring the magnetic fields given off by electrical currents (see Hari and Salmelin 2012 for an appraisal). This is not an exhaustive list: there are other ways of imaging the brain (such as near infrared spectroscopy [NIRS] or positron emission tomography [PET]), but for one reason or another, these were less well represented in the group of people interviewed for this book.

3   Steve Silberman (2015) has troubled this assumption in his lucid, wide-ranging book *Neurotribes*, which reveals that Kanner may have known more of Asperger's work than is generally thought. Silberman shows that Asperger's diagnostician Georg Frankl had actually joined Kanner's clinic in Baltimore when Kanner described the first autism patients who would go on to feature in his foundational 1943 paper.

4   A quick note on diagnostic manuals: Diagnostic manuals for psychiatric disorder, as attempts to systematize diagnosis and create some sense of validity around the diagnostic process, emerged in the 1950s but achieved real prominence during the 1980s. Probably the most widely known is the American Psychiatric Association's *Diagnostic and Statistical Manual of Mental Disorders*, widely known as the *DSM*, now in its fifth edition (known as *DSM-5*). This edition folded Asperger's and autism together. The *DSM* has partly achieved its prominence because its categories are typically used by researchers around the world; for reasons of standardization and comparability, they need to use the same diagnostic tools. But it is not the only game in town—certainly not in clinical usage and certainly not in the United Kingdom. The most well-known alternative is the World Health Organization's *International Classification of Disease (ICD)*, currently in its tenth edition. This manual, at the time of writing this manuscript, retains a diagnosis of Asperger's syndrome distinct from autism but is due an update in 2018 and has tended to track the *DSM*. An online draft (available at http://apps.who.int/classifications/icd11/browse/l-m/en) does indeed place the category of Asperger's syndrome within a new parent category of "Autism Spectrum Disorder," describing what *was* Asperger's syndrome as "Autism spectrum disorder without disorder of intellectual development and with mild or no impairment of functional language."

5   See Hollin 2014 for an important sociological account of cognitive theories of autism.

6   Such as, for example, differences in white matter and overall brain volume (see Herbert et al. 2004). More recently, researchers have moved away from interest in specific regions to a problem in connection between regions (Kleinhans et al. 2008). Others have worked on tools for grouping different neuroanatomical features together (Ecker et al. 2010) or have used a systems approach to match genetic heterogeneity to neurobiological pathophysiology (Willsey and State 2015).

7   Let me add here a short note on terminology. There is considerable debate about the use of terms such as *autism, autisms, autism spectrum condition(s), autism spectrum disorder(s)*, and so on. There are also debates around "person-first" language—which is the question of whether it's better to say "a person with autism" (to distinguish the person from the label) or, as some prefer, an "autistic person" or even "an autistic" (to insist that autism is a vital part of an individual's identity, not something to be wished away). See Kenny et al. 2015 for an interesting empirical analysis of these discussions. In what follows, I am not committing myself to a set terminology or a fixed terminological position; instead, I generally use words as they make sense within the (mostly psychological and research-based) context in which they appear. This means I use terms like "ASD" (for autism spectrum disorder) as well as sometimes simply "disorder." For the sake of legibility, I do not use scare quotes every time, and I rely on the broader spirit of the book to make it clear that, working through psychological accounts, I adopt that discipline's language as a shorthand only. I do not do so in ignorance of the politics and contest that still surround this language (see Singer 1999; Bumiller 2008).

8   I *don't* want to use this kind of language.

9   For example, we already have detailed attention to the complex somaticization and cerebralization of such diverse diagnostic categories as depression (Helén 2011), bipolar disorder (Martin 2007), psychopathy (Pickersgill 2009), addiction (Vrecko 2010), and even autism itself (Ortega and Choudhury 2009) as well as identifications of an emerging neurological basis to adolescence (Choudhury 2010), old age (Williams, Higgs, and Katz 2011), childhood difference (Rapp 2011), and human subjectivity in general (Ortega and Vidal 2007).

10 Here, and in subsequent sections of this introduction, I draw on and expand some arguments made in Fitzgerald and Callard 2015.

11 This useful gloss comes in the course of an essay in which Love (2010) is actually asserting her right to remain somewhat suspicious of such a claim, not least for what it might obscure in Sedgwick's (2003) own work.

12 Of these thirty-seven interviews, thirty were with autism neuroscientists, three with psychologists who worked on autism but not much in neuroscience, one with a geneticist, and three more with people from advocacy or charitable organizations. Overall, an initial trawl for potential interviewees produced a long list of sixty people. My search criteria were UK-based researchers who had a declared interest in, or record of publishing on, autism and who were either identifiably "neuroscientists" or whose work practice had a strongly neuroscientific component. I also included an additional number of non-neuroscientists who might have had a relevant view on this research. This long list included some neurogeneticists, biostatisticians, clinicians, and representatives from related fields. It also included a small number of representatives from what in the United Kingdom is sometimes called the "third sector" (i.e., not-for-profit

autism funding or advocacy organizations). Nonetheless, on the basis of their core training, publication history, or current declared interests, forty of these initial sixty individuals could be called "autism neuroscientists" without much fear of contradiction—and this was the group from which I ultimately interviewed my thirty autism neuroscientists. Those thirty interviews make up the vast bulk of the material quoted in the book. When I draw on an interview from one of the others, I make that interviewee's position clear.

13   This is an approach that many associate with "grounded theory," although I have no explicit affiliation to this approach. See Charmaz 2003 for a discussion.

14   See Gilbert and Mulkay 1984 for a classic of the genre within science and technology studies. See also Savage 2008 on the interviewing technique of Elizabeth Bott, which importantly historicizes the tradition of the (in-depth) interview in sociology, and indeed this suspicion of it, regarding its relations to the "psy" sciences.

15   My more explicit thoughts on this topic, which are all embedded in shared work with Felicity Callard, can be found in Fitzgerald and Callard 2015, and Callard and Fitzgerald 2015.

16   I am here partly motivated by the compelling work on "surface reading" that has emerged in studies of literature in the past few years (see Best and Marcus 2009), which has a strong family resemblance to the "nonparanoid" intervention I attempt here. Of course, there are other traditions of interviewing—many of them more interpretive and deep-diving—which I skate across here. See Savage 2008.

CHAPTER 1. THIS THING CALLED AUTISM

1   This term, which I use a few times in what follows, is generally associated with the anthropologist of medicine Annemarie Mol's (2003: ix, 33) excellent account of how multiply-real medical objects get "enacted" in practice. While there is some resonance with my project, I do not draw extensively on Mol, because my interest is not so much in the ontological whether or how of multiple real medical objects. Rather, I am interested in how a specific set of researchers talk through the connections between (1) diagnostic entities that are understood and apprehended in a multiplicity of ways and (2) a research practice that is nonetheless (at least formally) in pursuit of some kind of identifiable singularity. I focus very squarely on neuroscientists' shifting dynamics of contingency and stability to limn these accounts. I make no contribution to a broader claim for the ontology of medical objects (ibid.: 5).

2   See, for example, a monograph by Uta Frith (2003) (*Autism: Explaining the Enigma*); the blog of neuroscientist Jon Brock (2012) ("Cracking the Enigma: An Autism Research Blog"); and a special issue of *Nature* (2011) ("The Autism Enigma")—to choose only a few quick examples.

3   For research purposes, where standardization and comparability are crucial, specific cutoff points on the "gold standard" diagnostic scales are often used, such as the ADOS (Autism Diagnostic Observation Schedule) and ADI-R (Autism Diagnostic Interview—Revised).

CHAPTER 2. THE TROUBLE WITH BRAIN IMAGING

1   This ethic of excoriation has since been embroiled in the wider "replication crisis" in psychology (see Open Science Collaboration 2015), which is not about cognitive neuroscience as such, but through which nonetheless some of the more downbeat registers described here can and should be read. See also Eklund, Nichols, and Knutsson 2016.

2   Obviously there is some humor involved here; the descriptor "fucked" is mobilized in a very specific way. See Srivastava 2016.

3   Also known as the "default mode," this is data collected while the subject is supposed to be "at rest" in the scanner, not performing any specific function, but also in which, it later transpired, the brain was surprisingly active (Callard and Margulies 2011).

4   See the discussion of the HIV prophylactic pill known as PrEP in Rosengarten and Michael 2009 for an example of the way that expectations can form and reform around the changeable and emergent nature of scientific objects.

5   Indeed, see Pinch 2011 for a sense of the incredulity with which such a suggestion is greeted by a more conventional science and technology studies.

6   This discussion partly recapitulates a shorter account of these passages from Karen Barad, given in Fitzgerald and Callard 2015.

7   With different coauthors I have also written more extensively on the specific critical proposal embedded in the critical neuroscience literatures (see Fitzgerald et al. 2014). I hope that this extended attention, although founded on significant disagreement, signals the very high regard in which I hold the different forays that have emerged through "critical neuroscience" as well as the very important work that, in my view, this term has done and continues to do.

CHAPTER 3. THE THROBBING EMOTION OF THE PAST

1   I am grateful to an anonymous reviewer for drawing my attention to this aspect of the chapter.

2   Still less am I invested in the "wars" that have broken out around it. See, e.g., Leys 2011; Frank and Wilson 2012. For a critical take on the affective turn, with which I am in great sympathy, see Papoulias and Callard 2012.

3   I am grateful to an anonymous reviewer who noticed this before I did.

4   Here I am indebted to the argument that Elizabeth Wilson makes in *Gut Femi-*

nism (2015) about the energies produced by feminist critiques of biology and what happens politically when we start to think otherwise.

CHAPTER 4. COMPROMISING POSITIONS

1   See Rose and Abi-Rached (2013: 41–47) for a much more comprehensive account of the emergence of a "neuromolecular style of thought" across a range of disciplines during the twentieth century.
2   Max Bennett and Peter Hacker (2012: 2), for example, associate cognitive neuroscience with "the idea that synaptic networks in the brain possess psychological attributes."
3   Rapprochement with Freud is, of course, the focus of much neuropsychological attention elsewhere—see Hustvedt 2010 or Papoulias and Callard 2012 for discussions.
4   Insel has since left the National Institute of Mental Health for Google, which is surely its own (not unrelated) story.
5   See, for example, a wide-ranging review of gene-environment-interaction research on emotional and behavioral problems by Michael Rutter and Judy Silberg (2002: 478) that looks to "the operation of racial discrimination, availability of guns, local authority housing policies, availability of family planning, and schooling, to mention just a few examples" (Rutter, of course, is an absolutely foundational figure in the scientific study of autism).
6   To put it simply, for a long time the etiology of autism was associated with the interaction between mother and child—understood by many as imputing a sense of blame, to the mother in particular, for the onset of the disorder. Bruno Bettelheim (1967) is most infamously associated with this account, as more generally is the psychoanalytic paradigm with which Bettelheim associated himself. Both popped up frequently enough in my interviews but never in any depth, and almost always as loose metonyms for a dark and distant past.

CHAPTER 5. SEEING THE UNICORN

1   The public-facing NHS Choices website, for example, explains that "the exact cause of ASD is unknown, but it's thought several complex genetic and environmental factors are involved" (NHS Choices 2016).
2   Autism is often associated with unusual relationships to touch and to sensation in general (see Baranek et al. 2006; Grandin 2006).
3   "Aspie" is a familiar term for Asperger's syndrome, sometimes used by self-advocates.
4   In psychology, far more women than men have received doctorates for some decades (see, e.g., Willyard 2011). In neuroscience as such, the picture is more

mixed, with about half of all doctorates awarded to women, although it is likely too early to say much about the picture for neuroscience in general (see *Nature Neuroscience* 2006).

CONCLUSION

1   I am grateful to an anonymous reviewer, both for drawing my attention to this work and for suggesting this reading of it.

2   It's worth noting that there are other ways of figuring the historical relationship between critique and the history of feminist science studies—not least in the work of Donna Haraway herself. See Kenney and Müller 2016 for an alternative to the account I provide here.

BIBLIOGRAPHY

Abi-Rached, J. M. 2008. "The Implications of the New Brain Sciences." *EMBO Reports* 9 (12): 1158–62.

Abi-Rached, J. M., and N. Rose. 2010 "The Birth of the Neuromolecular Gaze." *History of the Human Sciences* 23 (1): 11–36.

Ahmed, S. 2004. *The Cultural Politics of Emotion.* Edinburgh: Edinburgh University Press.

Alaimo, S., and S. Hekman, eds. 2008. *Material Feminisms.* Bloomington: Indiana University Press.

Alleyne, R. 2010. "Brain Scan Could Diagnose Autism Early." *Telegraph.co.uk.* Available at www.telegraph.co.uk/health/healthnews/6951699/Brain-scan-could-diagnose-autism-early.html (accessed March 25, 2016).

American Psychiatric Association (APA). 2000. *Diagnostic and Statistical Manual of Mental Disorders: DSM-IV-TR.* Washington, DC: American Psychiatric Association Publishing.

———. 2013. *Diagnostic and Statistical Manual of Mental Disorders, Fifth Edition: DSM-5.* Washington, DC: American Psychiatric Association Publishing.

Anderson, J. S., T. J. Druzgal, A. Froehlich, M. B. DuBray, N. Lange, A. L. Alexander, T. Abildskov, J. A. Nielsen, A. N. Cariello, J. R. Cooperrider, E. D. Bigler, and J. E. Lainhart. 2011. "Decreased Interhemispheric Functional Connectivity in Autism." *Cerebral Cortex* 21 (5): 1134–46.

Andreasen, N. C. 2001. *Brave New Brain: Conquering Mental Illness in the Era of the Genome.* Oxford: Oxford University Press.

Arribas-Ayllon, M., A. Bartlett, and K. Featherstone. 2010. "Complexity and Accountability: The Witches' Brew of Psychiatric Genetics." *Social Studies of Science* 40 (4): 499–524.

Ash, M. G. 1992. "Historicizing Mind Science: Discourse, Practice, Subjectivity." *Science in Context* 5 (2): 193–207.

Baggs, A. 2007. *In My Language.* Available at www.youtube.com/watch?v=JnylM1hI2jc&feature=youtube_gdata_player (accessed June 1, 2012).

Bailey, A., A. Le Couteur, I. Gottesman, P. Bolton, E. Simonoff, E. Yuzda, and M. Rutter. 1995. "Autism as a Strongly Genetic Disorder: Evidence from a British Twin Study." *Psychological Medicine* 25 (1): 63–77.

Barad, K. 2007. *Meeting the Universe Halfway: Quantum Physics and the Entanglement of Matter and Meaning.* Durham, NC: Duke University Press.

———. 2011. "Erasers and Erasures: Pinch's Unfortunate 'Uncertainty Principle.'" *Social Studies of Science* 41 (3): 443–54.

Baranek, G. T., F. J. David, M. D. Poe, W. L. Stone, and L. R. Watson. 2006. "Sensory Experiences Questionnaire: Discriminating Sensory Features in Young Children with Autism, Developmental Delays, and Typical Development." *Journal of Child Psychology and Psychiatry* 47 (6): 591–601.

Barnbaum, D. R. 2008. *The Ethics of Autism: Among Them, but Not of Them.* Bloomington: Indiana University Press.

Barnes, E., and H. McCabe. 2012. "Should We Welcome a Cure for Autism? A Survey of the Arguments." *Medicine, Health Care, and Philosophy* 15 (3): 255–69.

Baron-Cohen, S. 1991. "Do People With Autism Understand What Causes Emotion?" *Child Development* 62 (2): 385–95.

———. 1995 *Mindblindness: An Essay on Autism and Theory of Mind.* Cambridge, MA: MIT Press.

Baron-Cohen, S., R. C. Knickmeyer, and M. K. Belmonte. 2005. "Sex Differences in the Brain: Implications for Explaining Autism." *Science* 310 (5749): 819–23.

Beaulieu, A. 2000. *The Space Inside the Skull: Digital Representations, Brain Mapping, and Cognitive Neuroscience in the Decade of the Brain.* University of Amsterdam.

———. 2002 "Images Are Not the (Only) Truth: Brain Mapping, Visual Knowledge, and Iconoclasm." Science, Technology, and Human Values 27 (1): 53–86.

Bennett, J. 2010. *Vibrant Matter: A Political Ecology of Things.* Durham, NC: Duke University Press.

Bennett, M. R., and P. M. Hacker. 2008. *History of Cognitive Neuroscience.* London: WileyBlackwell.

Best, S., and S. Marcus. 2009. "Surface Reading: An Introduction." *Representations* 108 (1): 1–21.

Betancur, C. 2011. "Etiological Heterogeneity in Autism Spectrum Disorders: More Than 100 Genetic and Genomic Disorders and Still Counting." *Brain Research* 1380: 42–77.

Bettelheim, B. 1967. *The Empty Fortress: Infantile Autism and the Birth of the Self.* New York: Free Press.

Blackman, Lucy. 2005. "Reflections on Language." In D. Biklen, ed., *Autism and the Myth of the Person Alone.* New York: New York University Press.

Blackman, Lisa. 2005. "The Dialogical Self, Flexibility, and the Cultural Production of Psychopathology." *Theory and Psychology* 15 (2): 183–206.

———. 2007. "Psychiatric Culture and Bodies of Resistance." *Body and Society* 13 (2): 1–23.

———. 2010. "Embodying Affect: Voice-hearing, Telepathy, Suggestion, and Modelling the Non-conscious." *Body and Society* 16 (1): 163–92.

Bleuler, E. 1951 [1913]. "Autistic Thinking." *American Journal of Insanity* 69 (5): 873–86.

Borck, C. 2008. "Recording the Brain at Work: The Visible, the Readable, and the

Invisible in Electroencephalography." *Journal of the History of the Neurosciences* 17 (3): 367–79.

Bor, D. 2012. "The Dilemma of Weak Neuroimaging Papers." *Daniel Bor: Author and Neuroscientist.* Available at www.danielbor.com/dilemma-weak-neuroimaging/ (accessed May 16, 2012).

Borup, M., N. Brown, K. Konrad, and H. Van Lente. 2006. "The Sociology of Expectations in Science and Technology." *Technology Analysis and Strategic Management* 18 (3–4): 285–98.

Brock, J. 2012. "Cracking the Enigma: An Autism Research Blog." *Cracking the Enigma: An Autism Research Blog.* Available at http://crackingtheenigma.blogspot.co.uk/ (accessed March 23, 2016).

Brown, N., and M. Michael. 2003. "A Sociology of Expectations: Retrospecting Prospects and Prospecting Retrospects." *Technology Analysis and Strategic Management* 15 (1): 3–18.

Brown, N., B. Rappert, and A. Webster. 2000. "Introducing Contested Futures: From Looking into the Future to Looking at the Future." In N. Brown, B. Rappert, and A. Webster, eds., *Contested Futures: A Sociology of Prospective Techno-Science.* Aldershot: Ashgate. Pp. 3–20.

Bumiller, K. 2008. "Quirky Citizens: Autism, Gender, and Reimagining Disability." *Signs* 33 (4): 967–91.

Butler, J. 1993. *Bodies That Matter: On the Discursive Limits of Sex.* London: Routledge.

Callard, F., and D. Fitzgerald. 2015. *Rethinking Interdisciplinarity across the Social Sciences and Neurosciences.* Basingstoke: Palgrave.

Callard, F., and D. S. Margulies. 2011. "The Subject at Rest: Novel Conceptualizations of Self and Brain from Cognitive Neuroscience's Study of the 'Resting State.'" *Subjectivity* 4 (3): 227–57.

Campbell, N. D. 2010. "Toward a Critical Neuroscience of 'Addiction.'" *BioSocieties* 5 (1): 89–104.

Canguilhem, G. 1980. "What Is Psychology." *Ideology and Consciousness* 7: 37–50.

Caspi, A., J. McClay, T. E. Moffitt, J. Mill, J. Martin, I. W. Craig, A. Taylor, and R. Poulton. 2002. "Role of Genotype in the Cycle of Violence in Maltreated Children." *Science* 297 (5582): 851–54.

Caspi, A., and T. E. Moffitt. 2006. "Gene-Environment Interactions in Psychiatry: Joining Forces with Neuroscience." *Nature Reviews Neuroscience* 7 (7): 583–90.

Centers for Disease Control (CDC). 2009. "Prevalence of Autism Spectrum Disorders—Autism and Developmental Disabilities Monitoring Network, United States, 2006." *Centers for Disease Control: Morbidity and Mortality Weekly Report* 58: 1–20.

———. 2014. "Prevalence of Autism Spectrum Disorder among Children Aged 8 Years—Autism and Developmental Disabilities Monitoring Network, 11 Sites, United States, 2010." *Centers for Disease Control: Morbidity and Mortality Weekly Report* 63: 1–21.

Chamak, B., B. Bonniau, E. Jaunay, and D. Cohen. 2008. "What Can We Learn About Autism from Autistic Persons?" *Psychotherapy and Psychosomatics* 77: 271–79.

Charmaz, K. 2003. "Qualitative Interviews and Grounded Theory Analysis" In J. A. Holstein and J. F. Gubrium, eds., *Inside Interviewing: New Lenses, New Concerns.* London: Sage. Pp. 311–30.

Choudhury, S. 2010. "Culturing the Adolescent Brain: What Can Neuroscience Learn from Anthropology?" *Social Cognitive and Affective Neuroscience* 5 (2–3): 159–67.

Choudhury, S., S. K. Nagel, and J. Slaby. 2009. "Critical Neuroscience: Linking Neuroscience and Society through Critical Practice." *BioSocieties* 4: 61–77.

Choudhury, S., and J. Slaby. 2011 "Introduction: Critical Neuroscience—Between Lifeworld and Laboratory" In S. Choudhury and J. Slaby, eds., *Critical Neuroscience: A Handbook of the Social and Cultural Contexts of Neuroscience.* London: Wiley-Blackwell. Pp. 1–26.

Clarke, A. E., L. Mamo, J. R. Fosket, J.R. Fishman, and J. K. Shim, eds. 2010. *Biomedicalization: Technoscience, Health, and Illness in the U.S.* Durham, NC: Duke University Press.

Clarke, A. E., J. K. Shim, L. Mamo, J. R. Fosket, and J. R. Fishman. 2003. "Biomedicalization: Technoscientific Transformations of Health, Illness, and U.S. Biomedicine." *American Sociological Review* 68 (2): 161–94.

Clough, P. T., and J. Halley, eds. 2007. *The Affective Turn: Theorizing the Social.* Durham, NC: Duke University Press.

Cohn, S. 2008. "Making Objective Facts from Intimate Relations: The Case of Neuroscience and Its Entanglements with Volunteers." *History of the Human Sciences* 21 (4): 86–103.

Coleman, M., and C. Gillberg. 2011. *The Autisms.* Oxford: Oxford University Press.

Collins, P. 2006. *Not Even Wrong: Adventures in Autism.* London: Bloomsbury.

Conrad, P. 1992. "Medicalization and Social Control." *Annual Review of Sociology* 18 (1): 209–32.

Daily Mail Reporter. 2011. "People Who Have Autistic Brother or Sister 'Carry Dormant Form of Disorder That Makes Them Less Empathetic.'" *The Daily Mail.* Available at www.dailymail.co.uk/health/article-2014192/Siblings-autism-sufferers-carry-dormant-form-disorder-affects-brain.html (accessed July 10, 2012).

Danziger, K. 1994. *Constructing the Subject: Historical Origins of Psychological Research.* Cambridge: Cambridge University Press.

Dapretto, M., M. S. Davies, J. H. Pfeifer, A. A. Scott, M. Sigman, S. Y. Bookheimer, and M. Iacoboni. 2006. "Understanding Emotions in Others: Mirror Neuron Dysfunction in Children with Autism Spectrum Disorders." *Nature Neuroscience* 9 (1): 28–30.

Daston, L., and P. Galison. 2007. *Objectivity.* New York: Zone Books.

Davidson, J. 2008. "Autistic Culture Online: Virtual Communication and Cultural Expression on the Spectrum." *Social and Cultural Geography* 9 (7): 791–806.

Dawson, M., I. Soulières, M. A. Gernsbacher, and L. Mottron. 2007. "The Level and Nature of Autistic Intelligence" *Psychological Science* 18 (8): 657–62.

Dehue, T. 2001. "Establishing the Experimenting Society. The Historical Origination of Social Experimentation According to the Randomized Controlled Design." *American Journal of Psychology* 114 (2): 283–302.

De Vos, J., and E. Pluth, eds. 2016. *Neuroscience and Critique: Exploring the Limits of the Neurological Turn.* London: Routledge.

Diener, E. 2010. "Neuroimaging: Voodoo, New Phrenology, or Scientific Breakthrough? Introduction to Special Section on fMRI." *Perspectives on Psychological Science* 5 (6): 714–15.

Donald, A. 2012. "End the Macho Culture That Turns Women Off Science." *The Guardian.* Available at www.guardian.co.uk/commentisfree/2012/jun/17/women-science-athene-donald (accessed June 22, 2012).

Dumit, J. 1999. "Objective Brains, Prejudicial Images." *Science in Context* 12 (1): 173–201.

———. 2004. *Picturing Personhood: Brain Scans and Biomedical Identity.* Princeton, NJ: Princeton University Press.

Ecker, C., A. Marquand, J. Mourao-Miranda, P. Johnston, E. M. Daly, M. J. Brammer, S. Maltezos, C. M. Murphy, D. Robertson, S. C. Williams, and D.G.M. Murphy. 2010. "Describing the Brain in Autism in Five Dimensions—Magnetic Resonance Imaging-Assisted Diagnosis of Autism Spectrum Disorder Using a Multiparameter Classification Approach." *Journal of Neuroscience* 30 (32): 10612–23.

Ehrenberg, A. 2011. "The 'Social' Brain: An Epistemological Chimera and a Sociological Truth" In F. Ortega and F. Vidal, eds., *Neurocultures: Glimpses into an Expanding Universe.* Frankfurt am Main: Peter Lang. Pp. 117–40.

Eklund, A., T. E. Nichols, and H. Knutsson. 2016 "Cluster Failure: Why fMRI Interfaces for Spatial Extent Have Inflated False-Positive Rates." *Proceedings of the National Academy of Sciences (PNAS)* 13 (28): 7900–905

Evans, B. 2017. *The Metamorphosis of Autism: A History of Child Development in Britain.* Manchester: Manchester University.

Eyal, G., D. Fitzgerald, E. Gillis-Buck, B. Hart, M. D. Lappé, D. Navon, and S. S. Richardson. 2014. "New Modes of Understanding and Acting on Human Difference in Autism Research, Advocacy, and Care: Introduction to a Special Issue of BioSocieties." *BioSocieties* 9 (3): 233–40.

Eyal, G., and B. Hart. 2010. "How Parents of Autistic Children Became Experts on Their Own Children: Notes towards a Sociology of Expertise" *Annual Conference of the Berkeley Journal of Sociology.* Available at http://works.bepress.com/cgi/viewcontent.cgi?article=1000&context=gil_eyal (accessed July 19, 2012).

Eyal, G., B. Hart, E. Onculer, N. Oren, and N. Rossi. 2011. *The Autism Matrix: The Social Origins of the Autism Epidemic.* London: Polity.

Farah, M. 2005. "Neuroethics: The Practical and the Philosophical." *Neuroethics Publications.* Available at http://repository.upenn.edu/neuroethics_pubs/8.

Feinstein, A. 2010. *A History of Autism: Conversations with the Pioneers.* London: Wiley-Blackwell.

Felski, R. 2015. *The Limits of Critique.* Chicago: University of Chicago Press.

Fine, E. 2016. "Our Circuits, Ourselves: What the Autism Spectrum Can Tell Us about the Research Domain Criteria Project (RDoC) and the Neurogenetic Transformation of Diagnosis." *BioSocieties* 11 (2): 175–98.

Fine, S. E., A. Weissman, M. Gerdes, J. Pinto-Martin, E. H. Zackai, D. M. McDonald-McGinn, and B. S. Emanuel. 2005. "Autism Spectrum Disorders and Symptoms in Children with Molecularly Confirmed 22q11.2 Deletion Syndrome." *Journal of Autism and Developmental Disorders* 35 (4): 461–70.

Fitzgerald, D., and F. Callard. 2015. "Social Science and Neuroscience beyond Interdisciplinarity: Experimental Entanglements." *Theory, Culture, and Society* 32 (1): 3–32.

Fitzgerald, D., S. Matusall, J. Skewes, and A. Roepstorff. 2014. "What's So Critical about Critical Neuroscience? Rethinking Experiment, Enacting Critique." *Frontiers in Human Neuroscience* 8, art. 365.

Fitzgerald, D., N. Rose, and I. Singh. 2016. "Revitalizing Sociology: Urban Life and Mental Illness between History and the Present." *British Journal of Sociology* 67 (1): 138–60.

Folkers, A. 2016. "Daring the Truth: Foucault, Parrhesia, and the Genealogy of Critique." *Theory, Culture, and Society* 33 (1): 3–28: 0263276414558885.

Fortun, M. 2008. *Promising Genomics: Iceland and DeCODE Genetics in a World of Speculation.* Berkeley: University of California Press.

Fox Keller, E. 1977. "The Anomaly of a Woman in Physics" In S. Ruddick and P. Daniels, eds., *Working It Out.* New York: Pantheon Books. Pp. 71–91.

Frank, A., and E. A. Wilson. 2012. "Like-Minded." *Critical Inquiry* 38 (4): 870–77.

Frith, C. D. 2007. *Making up the Mind: How the Brain Creates Our Mental World.* Oxford: Blackwell Publishing.

Frith, U. 2003. *Autism: Explaining the Enigma.* 2nd edition. London: Wiley-Blackwell.

———. 2008. *Autism: A Very Short Introduction.* Oxford: Oxford University Press.

Frith, U., and F. Happé. 1994. "Autism: Beyond 'theory of mind.'" *Cognition* 50: 115–32.

Fuchs, A. H., and K. S. Milar. 2003. "Psychology as a Science." In *Handbook of Psychology.* London: Wiley. Available at http://onlinelibrary.wiley.com/doi/10.1002/0471264385.wei0101/abstract (accessed March 25, 2016).

Gardner, J., G. Samuel, and C. Williams. 2015. "Sociology of Low Expectations Recalibration as Innovation Work in Biomedicine." *Science, Technology, and Human Values* 40 (6): 998–1021.

Gaudion, K., A. Hall, J. Myerson, and L. Pellicano. 2015. "A Designer's Approach: How Can Autistic Adults with Learning Disabilities Be Involved in the Design Process?" *CoDesign* 11 (1): 49–69.

Gernsbacher, M. A., M. Dawson, and L. Mottron. 2006. "Autism: Common, Heritable but Not Harmful." *Behavioral and Brain Sciences* 29 (4): 413–14.

Geschwind, D. H. 2009. "Advances in Autism." *Annual Review of Medicine* 60 (1): 367–80.

Geschwind, D. H., and P. Levitt. 2007. "Autism Spectrum Disorders: Developmental Disconnection Syndromes." *Current Opinion in Neurobiology* 17 (1): 103–11.

Gieryn, T. F. 1983. "Boundary-Work and the Demarcation of Science from Non-Science: Strains and Interests in Professional Ideologies of Scientists." *American Sociological Review* 48 (6): 781–95.

Gilbert, G. N., and M. Mulkay. 1984. *Opening Pandora's Box: A Sociological Analysis of Scientists' Discourse.* Cambridge: Cambridge University Press.

Gillies, V., R. Edwards, and N. Horsley. 2016. "Brave New Brains: Sociology, Family, and the Politics of Knowledge." *The Sociological Review* 62 (2): 219–37.

Gordon, D. 1988. "Clinical Science And Clinical Expertise: Changing Boundaries Between Art and Science in Medicine" In M. Lock and D. Gordon, eds., *Biomedicine Examined.* London: Kluwer Academic Publishers. Pp. 257–95.

Grandin, T. 2005. *Emergence: Labelled Autistic.* New York: Grand Central Publishing.

———. 2006. *Thinking in Pictures.* London: Bloomsbury.

Grandin, T., and R. Panek. 2014. *The Autistic Brain: Helping Different Kinds of Brain Succeed.* Boston: Mariner.

Gregg, M., and G. J. Seigworth, eds. 2010. *The Affect Theory Reader.* Durham, NC: Duke University Press.

Grinker, R. R. 2007. *Unstrange Minds: Remapping the World of Autism.* New York: Basic Books.

Grosz, E. A. 1994. *Volatile Bodies: Toward a Corporeal Feminism.* Bloomington: Indiana University Press.

Gupta, A. R., and M. W. State. 2007. "Recent Advances in the Genetics of Autism." *Biological Psychiatry* 61 (4): 429–37.

Hacking, I. 2006a. "Making up People." *London Review of Books* 28 (16). Available at www.lrb.co.uk/v28/n16/ian-hacking/making-up-people (accessed November 3, 2016).

———. 2006b. "What Is Tom Saying to Maureen?." *London Review of Books* 28 (9): 3–7.

Happé, F. 1996. "Studying Weak Central Coherence at Low Levels: Children with Autism Do Not Succumb to Visual Illusions. A Research Note." *Journal of Child Psychology and Psychiatry* 37 (7): 873–77.

———. 1999. "Autism: Cognitive Deficit or Cognitive Style?" *Trends in Cognitive Sciences* 3 (6): 216–22.

Happé, F., and A. Ronald. 2008. "The 'Fractionable Autism Triad': A Review of Evidence from Behavioural, Genetic, Cognitive, and Neural Research." *Neuropsychology Review* 18 (4): 287–304.

Happé, F., A. Ronald, and R. Plomin. 2006. "Time to Give Up on a Single Explanation for Autism." *Nature Neuroscience* 9 (10): 1218–20.

Haraway, D. J. 1985. "Manifesto for Cyborgs: Science, Technology, and Socialist Feminism in the 1980s." *Socialist Review* 89: 65–108.

———. 1988. "Situated Knowledges: The Science Question in Feminism and the Privilege of Partial Perspective." *Feminist Studies* 14 (3): 575–99.

———. 1990. *Simians, Cyborgs, and Women: The Re-Invention of Nature.* New York: London: Free Association.

———. 1997. *Modest_Witness@Second_Millennium.FemaleMan_Meets_OncoMouse: Feminism and Technoscience.* London: Routledge.

———. 2007. *When Species Meet.* Minneapolis: University of Minnesota Press.

———. 2016. *Staying with the Trouble: Making Kin in the Chthuluscene.* Durham, NC: Duke University Press.

Hari, R., and R. Salmelin. 2012. "Magnetoencephalography: From SQUIDs to Neuroscience: Neuroimage 20th Anniversary Special Edition." *NeuroImage* 61 (2): 386–96.

Harman, G. 2009. *Prince of Networks: Bruno Latour and Metaphysics.* Melbourne: re.press.

Harrington, A. 2005. "The Inner Lives of Disordered Brains." *Cerebrum* (online magazine of the Dana Foundation). Available at www.dana.org/news/cerebrum/detail.aspx?id=794 (accessed June 20, 2012).

Hart, B. 2014. "Autism Parents and Neurodiversity: Radical Translation, Joint Embodiment, and the Prosthetic Environment." *BioSocieties* 9 (3): 284–303.

Hayles, K. 1999. *How We Became Posthuman: Virtual Bodies in Cybernetics, Literature, and Informatics.* Chicago: University of Chicago Press.

Hedgecoe, A. 1998. "Geneticization, Medicalisation, and Polemics." *Medicine, Health Care, and Philosophy* 1 (3): 235–43.

Helén, I. 2011. "The Depression Paradigm and Beyond: The Practical Ontology of Mood Disorders." *Science Studies* 24 (1): 81–112.

Herbert, M. R., D. A. Ziegler, N. Makris, P. A. Filipek, T. L. Kemper, J. J. Normandin, H. A. Sanders, D. N. Kennedy, and V. A. Caviness. 2004. "Localization of White Matter Volume Increase in Autism and Developmental Language Disorder." *Annals of Neurology* 55 (4): 530–40.

Hird, M. J., and C. Roberts. 2011. "Feminism Theorises The Nonhuman." *Feminist Theory* 12 (2): 109–17.

Hobson, R. P. 2011. "On the Relations between Autism and Psychoanalytic Thought and Practice." *Psychoanalytic Psychotherapy* 25 (3): 229–44.

Hollin, G. 2014. "Constructing a Social Subject: Autism and Human Sociality in the 1980s." *History of the Human Sciences* 27 (4): 98–115.

Houston, R., and U. Frith. 2000. *Autism in History: The Case of Hugh Blair of Borgue.* London: WileyBlackwell.

Howlin, P., S. Goode, J. Hutton, and M. Rutter. 2004. "Adult Outcome for Children with Autism." *Journal of Child Psychology and Psychiatry* 45 (2): 212–29.

Hustvedt, S. 2010. "Who's Afraid of Sigmund Freud?." Available at www.psychologytoday.com/blog/reading-minds-method-or-muddle/201003/who-s-afraid-sigmund-freud (accessed March 25, 2016).

Hyman, S. E. 2008. "A Glimmer of Light for Neuropsychiatric Disorders." *Nature* 455 (7215): 890–93.

———. 2009. "How Adversity Gets under the Skin." *Nature Neuroscience* 12 (3): 241–43.

Insel, T. 2010. "Faulty Circuits." *Scientific American* 302 (4): 44–51.

———. 2014. "The NIMH Research Domain Criteria (RDoC) Project: Precision Medicine for Psychiatry." *American Journal of Psychiatry* 171 (4): 395–97.

Jaarsma, P., and S. Welin. 2012. "Autism as a Natural Human Variation: Reflections on the Claims of the Neurodiversity Movement." *Health Care Analysis* 20 (1): 20–30.

Johnson, J., and M. Littlefield. 2011. "Lost and Found in Translation: Popular Neuroscience and the Emergent Neurodisciplines." *Sociological Reflections on Neuroscience* 13: 279–97.

Jones, E. G., and L. M. Mendell. 1999. "Assessing the Decade of the Brain." *Science* 284 (5415): 739–39.

Joyce, K. 2005. "Appealing Images Magnetic Resonance Imaging and the Production of Authoritative Knowledge." *Social Studies of Science* 35 (3): 437–62.

———. 2008. *Magnetic Appeal: MRI and the Myth of Transparency.* Ithaca, NY: Cornell University Press.

Just, M. A., V. L. Cherkassky, T. A. Keller, R. K. Kana, and N. J. Minshew. 2007. "Functional and Anatomical Cortical Underconnectivity in Autism: Evidence from an fMRI Study of an Executive Function Task and Corpus Callosum Morphometry." *Cerebral Cortex* 17 (4): 951–61.

Kandel, E. R. 2007. *In Search of Memory: The Emergence of a New Science of Mind.* New York: W. W. Norton & Co.

Kanner, L. 1949. "Problems of Nosology and Psychodynamics of Early Infantile Autism." *American Journal of Orthopsychiatry* 19 (3): 416–26.

———. 1968 [1943]. "Autistic Disturbances of Affective Contact." *Acta Paedopsychiatrica* 35 (4–8): 100–36.

Kapur, S., A. G. Phillips, and T. R. Insel. 2012. "Why Has It Taken So Long for Biological Psychiatry To Develop Clinical Tests and What To Do About It?." *Molecular Psychiatry* 17 (12): 1174–79.

Kenney, M., and R. Müller. 2016. "Of Rats and Mothers: Narratives of Motherhood in Environmental Epigenetics." *BioSocieties* online First. doi:10.1057/s41292-016-0002-7 (accessed November 3, 2016).

Kenny, L., C. Hattersley, B. Molins, C. Buckley, C. Povey, and E. Pellicano. 2015. "Which Terms Should Be Used To Describe Autism? Perspectives from the UK Autism Community." *Autism* 20 (4): 442–62.

King, M. D., and P. Bearman. 2011. "Socioeconomic Status and the Increased Prevalence of Autism in California." *American Sociological Review* 76 (2): 320–46.

Kirmayer, L. J. 2011. "The Future of Critical Neuroscience" In S. Choudhury and J. Slaby, eds., *Critical Neuroscience: A Handbook of the Social and Cultural Contexts of Neuroscience.* London: Wiley Blackwell. Pp. 367–83.

Kitzinger, J. 2008. "Questioning Hype, Rescuing Hope? The Hwang Stem Cell Scandal and the Reassertion of Hopeful Horizons." *Science as Culture* 17 (4): 417–34.

Kleinhans, N. M., T. Richards, L. Sterling, K. C. Stegbauer, R. Mahurin, L. C. Johnson, J. Greenson, G. Dawson, and E. Aylward. 2008. "Abnormal Functional Connectivity in Autism Spectrum Disorders during Face Processing." *Brain* 131 (4): 1000–12.

Klin, A. 2002. "Defining and Quantifying the Social Phenotype in Autism." *American Journal of Psychiatry* 159 (6): 895–908.

Knorr Cetina, K. 1999. *Epistemic Cultures: How the Sciences Make Knowledge*. Cambridge, MA: Harvard University Press.

Langlitz, N. 2010. "The Persistence of the Subjective in Neuropsychopharmacology: Observations of Contemporary Hallucinogen Research." *History of the Human Sciences* 23 (1): 37–57.

Lappé, M. 2014. "Taking Care: Anticipation, Extraction, and the Politics of Temporality in Autism Science." *BioSocieties* 9 (3): 304–28.

Latour, B. 1987. *Science in Action: How To Follow Scientists and Engineers through Society*. Cambridge: Harvard University Press.

———. 1993. *We Have Never Been Modern*. Cambridge: Harvard University Press.

———. 1996. *Aramis, or the Love of Technology*. Cambridge: Harvard University Press.

———. 2004. "Why Has Critique Run out of Steam? From Matters of Fact to Matters of Concern." *Critical Inquiry* 30 (2): 225–48.

———. 2008. *What Is the Style of Matters of Concern? Two Lectures in Empirical Philosophy*. Amsterdam: Van Gorcum.

———. 2012. "Paris, Invisible City: The Plasma." *City, Culture and Society*. Available at www.sciencedirect.com/science/article/pii/S1877916611000737 (accessed August 11, 2012).

Law, J. 2002. "Objects and Spaces." *Theory, Culture, and Society* 19 (5–6): 91–105.

Lecourt, D. 1975. *Marxism and Epistemology: Bachelard, Canguilhem, and Foucault*. London: New Left Books.

Leys, B. R. 2011. "The Turn to Affect: A Critique." *Critical Inquiry* 37 (3): 434–72.

Littlewood, R., and M. Lipsedge. 1997. *Aliens and Alienists: Ethnic Minorities and Psychiatry*. London: Routledge.

Lock, M. 2013. *The Alzheimer Conundrum: Entanglements of Dementia and Aging*. Princeton: Princeton University Press.

Logothetis, N. K. 2008. "What We Can Do and What We Cannot Do with fMRI." *Nature* 453 (7197): 869–78.

Lord, C., and R. M. Jones. 2012. "Annual Research Review: Re-thinking the Classification of Autism Spectrum Disorders." *Journal of Child Psychology and Psychiatry* 53 (5): 490–509.

Love, H. 2010. "Truth and Consequences: On Paranoid Reading and Reparative Reading." *Criticism* 52 (2): 235–41.

Lowe, P., E. Lee, and J. Macvarish. 2015. "Biologising Parenting: Neuroscience

Discourse, English Social, and Public Health Policy and Understandings of the Child." *Sociology of Health and Illness* 37 (2): 198–211.

Luhrmann, T. M. 2001. *Of Two Minds: An Anthropologist Looks at American Psychiatry.* New York: Vintage.

Lupton, D. 1997. "Foucault and the Medicalisation Critique." In A. Petersen and R. Bunton, eds., *Foucault, Health, and Medicine.* London: Psychology Press. Pp. 94–110.

Lynch, M. 1985. *Art and Artifact in Laboratory Science: A Study of Shop Work and Shop Talk in a Research Laboratory.* London: Routledge & Kegan Paul.

Mackenzie, B. D. 1977. *Behaviourism and the Limits Of Scientific Method.* London: Routledge.

Malterud, K. 2001. "The Art and Science of Clinical Knowledge: Evidence Beyond Measures and Numbers." *The Lancet* 358 (9279): 397–400.

Martin, E. 2000. "AES Presidential Address—Mind-Body Problems." *American Ethnologist* 27(3): 569–90.

———. 2004. "Talking Back To Neuro-Reductionism." In H. Thomas and J. Ahmed, eds., *Cultural Bodies: Ethnography and Theory.* Oxford: Blackwell. Pp. 190–211.

———. 2007. *Bipolar Expeditions: Mania and Depression in American Culture.* Princeton: Princeton University Press.

Massumi, B. 2002. *Parables for the Virtual: Movement, Affect, Sensation.* Durham, NC: Duke University Press.

Meillassoux, Q. 2008. *After Finitude: An Essay on the Necessity of Contingency.* London: Continuum.

Meloni, M. 2014. "The Social Brain Meets the Reactive Genome: Neuroscience, Epigenetics, and the New Social Biology." *Frontiers in Human Neuroscience* 8: 309.

Merton, R. 1979. *The Sociology of Science: Theoretical and Empirical Investigations.* Chicago: University of Chicago Press.

Mines, M. A., C. J. Yuskaitis, M. K. King, E. Beurel, and R. S. Jope. 2010. "GSK3 Influences Social Preference and Anxiety-Related Behaviors during Social Interaction in a Mouse Model of Fragile X Syndrome and Autism." *PloS One* 5 (3): e9706.

Mol, A. 2003. *The Body Multiple: Ontology in Medical Practice.* Durham, NC: Duke University Press.

———. 2008. *The Logic of Care: Health and the Problem of Patient Choice.* London: Routledge.

Moreira, T., and P. Palladino. 2005. "Between Truth and Hope: On Parkinson's Disease, Neurotransplantation and the Production of the 'Self.'" *History of the Human Sciences* 18 (3): 55–82.

Morton, J. 2004. *Understanding Developmental Disorders: A Causal Modelling Approach.* London: Wiley-Blackwell.

Murray, S. 2008. *Representing Autism: Culture, Narrative, Fascination.* Liverpool: Liverpool University Press.

———. 2011. *Autism.* London: Routledge.

Myers, N. 2012. "Dance Your PhD: Embodied Animations, Body Experiments, and the Affective Entanglements of Life Science Research." *Body and Society* 18 (1): 151–89.

Nadesan, M. H. 2005. *Constructing Autism.* London: Routledge.

Nature. 2010. "A Decade for Psychiatric Disorders." *Nature* 463 (7277): 9.

———. 2011. "The Autism Enigma." *Nature* 479 (7371): 21.

Nature Neuroscience. 2006. "Women in Neuroscience: A Numbers Game." *Nature Neuroscience* 9 (7): 853.

Navon, D., and G. Eyal. 2016. "Looping Genomes: Diagnostic Change and the Genetic Makeup of the Autism Population." *American Journal of Sociology* 121 (5): 1416–71.

Nerlich, B., and C. Halliday. 2007. "Avian Flu: The Creation of Expectations in the Interplay between Science and the Media." *Sociology of Health and Illness* 29 (1): 46–65.

Neurocritic. 2012. "How Much of the Neuroimaging Literature Should We Discard?" *The Neurocritic: Deconstructing the Most Sensationalistic Recent Findings in Human Brain Imaging, Cognitive Neuroscience, and Psychopharmacology.* Available at http://neurocritic.blogspot.co.uk/2012/03/how-much-of-neuroimaging-literature.html (accessed March 23, 2016).

NHS Choices. 2016. "Causes of Autism Spectrum Disorder." *NHS Choices.* Available at www.nhs.uk/Conditions/Autistic-spectrum-disorder/Pages/Causes.aspx (accessed August 2016).

Niewöhner, J. 2011. "Epigenetics: Embedded Bodies and the Molecularisation of Biography and Milieu." *BioSocieties* 6: 279–98.

Oakley, A. 2016. "Interviewing Women Again: Power, Time, and the Gift." *Sociology* 50 (1): 195–213.

Ochs, E., and O. Solomon. 2010. "Autistic Sociality." *Ethos* 38 (1): 69–92.

Open Science Collaboration. 2015. "Estimating the Reproducibility of Psychological Science." *Science* 349 (6251): aac4716.

Orr, J. 2010. "Biopsychiatry and the Informatics of Diagnosis: Governing Mentalities" In A. E. Clarke, L. Mamo, J. R. Fosket, J. R. Fishman, and J. K. Shim, eds., *Biomedicalization: Technoscience, Health, and Illness in the U.S.* Durham, NC: Duke University Press. Pp. 353–79.

Orsini, M. 2012. "Autism, Neurodiversity, and the Welfare State: The Challenges of Accommodating Neurological Difference." *Canadian Journal of Political Science/ Revue canadienne de science politique* 45 (4): 805–27.

Ortega, F. 2009. "The Cerebral Subject and the Challenge of Neurodiversity." *BioSocieties* (4): 425–45.

Ortega, F., and S. Choudhury. 2011. "'Wired Up Differently': Autism, Adolescence, and the Politics of Neurological Identities." *Subjectivity* 4 (3): 323–45.

Ortega, F., and F. Vidal. 2007. "Mapping the Cerebral Subject in Contemporary Culture." *RECIIS: Electronic Journal of Communication Information and Innovation in Health* 1 (2): 255–59.

Osteen, M., ed. 2007. *Autism and Representation*. London: Routledge.

Oxford English Dictionary. 2016a. "enigma, n." *OED Online*. Available at http://oed
.com/view/Entry/62382?redirectedFrom=enigma#eid (accessed March 2016).

———. 2016b. "trace, n.1." *OED Online*. Available at http://oed.com/view/Entry/20417
4?rskey=O49UQD&result=1&isAdvanced=false#eid (accessed August 2016).

———. 2016c. "trace, v.1." *OED Online*. Available at http://oed.com/view/Entry/20417
7?rskey=U83cZl&result=4&isAdvanced=false (accessed March 2016).

Papoulias, C., and F. Callard. 2010. "Biology's Gift: Interrogating the Turn to
Affect." *Body and Society* 16 (1): 29–56.

———. 2012. "The Rehabilitation of the Drive in Neuropsychoanalysis: From Sexu-
ality to Self-Preservation." In *Freuds Referenzen*. Berlin: Kadmos. Available at
http://ssl.einsnull.com/paymate/search.php?vid=5&aid=3303 (accessed March 23,
2016).

Park, C. C. 1982 [1967]. *The Siege: A Family's Journey into the World of an Autistic Child*.
Boston: Back Bay Books.

Peart, K. 2015. "Yale Leads NIH-Funded Autism Biomarkers Study of Pre-School and
School-Aged Children." *Yale News*. Available at http://news.yale.edu/2015/07/20/
yale-leads-nih-funded-autism-biomarkers-study-pre-school-and-school-aged
-children (accessed March 21, 2016).

Pellicano, E., A. Dinsmore, and T. Charman. 2014. "What Should Autism Research
Focus Upon? Community Views and Priorities from the United Kingdom."
Autism 18 (7): 756–70.

Pellicano, E., A. Ne'eman, and M. Stears. 2011. "Engaging, Not Excluding: A
Response To Walsh et al." *Nature Reviews Neuroscience* 12 (12): 769–69.

Persico, A. M., and T. Bourgeron. 2006. "Searching for Ways Out of the Autism
Maze: Genetic, Epigenetic, and Environmental Clues." *Trends in Neurosciences* 29
(7): 349–58.

Pickersgill, M. 2009. "Between Soma and Society: Neuroscience and the Ontology
of Psychopathy." *BioSocieties* 4 (1): 45–60.

———. 2011a. "Neurogenetic Diagnoses: The Power of Hope, and the Limits of
Today's Medicine." *New Genetics and Society* 30 (1): 133–35.

———. 2011b. "'Promising' Therapies: Neuroscience, Clinical Practice, and the
Treatment of Psychopathy." *Sociology of Health and Illness* 33 (3): 448–64.

Pinch, T. 2011. "Review Essay: Karen Barad, Quantum Mechanics, and the Paradox
of Mutual Exclusivity." *Social Studies of Science* 41 (3): 431–41.

Pitts-Taylor, V. 2016. *The Brain's Body: Neuroscience and Corporeal Politics*. Durham, NC:
Duke University Press.

Porter, T. M. 1996. *Trust in Numbers: The Pursuit of Objectivity in Science and Public Life*.
Princeton, NJ: Princeton University Press.

Prince-Hughes, D. 2005. *Songs of the Gorilla Nation: My Journey through Autism*. New
York: Three Rivers Press.

Raff, M. 2009. "New Routes into the Human Brain." *Cell* 139 (7): 1209–11.

Raichle, M. E., A. M. MacLeod, A. Z. Snyder, W. J. Powers, D. A. Gusnard, and G. L. Shulman. 2001. "A Default Mode of Brain Function." *Proceedings of the National Academy of Sciences* 98 (2): 676–82.

Rapp, R. 2010. "Chasing Science: Children's Brains, Scientific Inquiries, and Family Labors." *Science, Technology, and Human Values* 36 (5): 662–84.

———. 2011. "A Child Surrounds This Brain: The Future of Neurological Difference According To Scientists, Parents, and Diagnosed Young Adults." *Advances in Medical Sociology* 13: 3–26.

Renwick. 2012. *British Sociology's Lost Biological Roots: A History of Futures Past.* Basingstoke: Palgrave.

Ring, H., M. Woodbury-Smith, P. Watson, S. Wheelwright, and S. Baron-Cohen. 2008. "Clinical Heterogeneity among People with High Functioning Autism Spectrum Conditions: Evidence Favouring a Continuous Severity Gradient." *Behavioral and Brain Functions* 4 (11). Available at www.repository.cam.ac.uk/bitstream/handle/1810/237986/1744-9081-4-11.pdf?sequence=2&isAllowed=y.

Roberts, T.P.L., S. Y. Khan, M. Rey, J. F. Monroe, K. Cannon, L. Blaskey, S. Woldoff, S. Qasmieh, M. Gandal, G. L. Schmidt, D. M. Zarnow, S. E. Levy, and J. C. Edgar. 2010. "MEG Detection of Delayed Auditory Evoked Responses in Autism Spectrum Disorders: Towards an Imaging Biomarker for Autism." *Autism Research* 3 (1): 8–18.

Robertson, S., and A. Ne'eman. 2008. "Autistic Acceptance, the College Campus, and Technology: The Growth of Neurodiversity in Society and Academia." *Disability Studies Quarterly* 28 (4). Available at http://dsq-sds.org/article/view/146/146 (accessed November 3, 2016).

Roepstorff, A. 2004. "Mapping Brain Mappers: An Ethnographic Coda." In R.S.J. Frackowiak, J. T. Ashburner, W. D. Penny, and S. Zeki, eds., *Human Brain Function*. London: Elsevier. Pp. 1105–17.

Ronald, A., F. Happé, P. Bolton, L. M. Butcher, T. S. Price, S. Wheelwright, S. Baron-Cohen, and R. Plomin. 2006. "Genetic Heterogeneity Between the Three Components of the Autism Spectrum: A Twin Study." *Journal of the American Academy of Child and Adolescent Psychiatry* 45 (6): 691–99.

Rose, H. 1994. *Love, Power, and Knowledge: Towards a Feminist Transformation of the Sciences.* London: Polity Press.

Rose, N. 1985. *The Psychological Complex: Psychology, Politics, and Society in England, 1869–1939.* London: Routledge Kegan & Paul.

———. 1996a. *Inventing Our Selves: Psychology, Power, and Personhood.* Cambridge: Cambridge University Press.

———. 1996b. "Power and Subjectivity: Critical History and Psychology" In C. F. Graumann and K. J. Gergen, eds., *Historical Dimensions of Psychological Discourse.* Cambridge: Cambridge University Press. Pp. 103–24.

———. 1999. *Governing the Soul: The Shaping of the Private Self.* London: Free association.

———. 2001. "The Neurochemical Self and Its Anomalies (pre-print)." Available at http://www2.lse.ac.uk/sociology/pdf/Rose-TheNeurochemicalSelfandItsAnomaliesOct01.pdf (accessed September 23, 2012).

———. 2007. *Politics of Life Itself: Biomedicine, Power, and Subjectivity in the Twenty-First Century.* Princeton, NJ: Princeton University Press.

———. 2013. "The Human Sciences in a Biological Age." *Theory, Culture, and Society* 30 (1): 3–34.

Rose, N., and J. M. Abi-Rached. 2013. *Neuro: The New Brain Sciences and the Management of the Mind.* Princeton, NJ: Princeton University Press.

Rose, S. 2004. "Introduction: The New Brain Sciences" In D. Rees and S. Rose, eds., *The New Brain Sciences: Perils and Prospects.* Cambridge: Cambridge University Press. Pp. 3–14.

Rosengarten, M., and M. Michael. 2009. "The Performative Function of Expectations in Translating Treatment to Prevention: The Case of HIV Pre-Exposure Prophylaxis, or PrEP." *Social Science and Medicine* 69 (7): 1049–55.

Rutter, M. 1968. "Concepts of Autism: A Review of Research." *Journal of Child Psychology and Psychiatry, and Allied Disciplines* 9 (1): 1–25.

———. 2005. "Environmentally Mediated Risks for Psychopathology: Research Strategies and Findings." *Journal of the American Academy of Child and Adolescent Psychiatry* 44 (1): 3–18.

Rutter, M., and J. Silberg. 2002 "Gene-Environment Interplay in Relation to Emotional and Behavioral Disturbance." *Annual Review of Psychology* 53 (1): 463–90.

Sacks, O. 1995. *An Anthropologist on Mars.* New York: Picador.

Savage, M. 2008. "Elizabeth Bott and the Formation of Modern British Sociology." *The Sociological Review* 56 (4): 579–605.

Schreibman, L. 2005. *The Science and Fiction of Autism.* Cambridge, MA: Harvard University Press.

Sedgwick, E. K. 2003. *Touching Feeling: Affect, Pedagogy, Performativity.* Durham, NC: Duke University Press.

Serres, M. 2007. *Parasite.* Minneapolis: University of Minnesota Press.

Shakespeare, T., and N. Watson. 2001. "The Social Model of Disability: An Outdated Ideology?" In S. N. Barnartt and B. M. Altman, eds., *Exploring Theories and Expanding Methodologies: Where We Are and Where We Need To Go (Research in Social Science and Disability, Volume 2).* Bingley, UK: Emerald. Pp. 9–28.

Shapin, S. 2010. *The Scientific Life: A Moral History of a Late Modern Vocation.* Chicago: University of Chicago Press.

Shaviro, S. 2009. "Pulses of Emotion: Whitehead's 'Critique of Pure Feeling.'" Available at www.shaviro.com/Othertexts/Pulse.pdf (accessed August 2016).

Silberman, S. 2015. *Neurotribes: The Legacy of Autism and How to Think Smarter about People Who Think Differently.* London: Allen & Unwin.

Silverman, C. 2008. "Critical Review: Fieldwork on Another Planet: Social Science Perspectives on the Autism Spectrum." *BioSocieties* 3: 325–41.

———. 2011. *Understanding Autism: Parents, Doctors, and the History of a Disorder.* Princeton, NJ: Princeton University Press.

Sinclair, J. 2013. "Why I Dislike 'Person-First' Language." *Autonomy, the Critical Journal of Interdisciplinary Autism Studies* 1 (2). Available at www.larry-arnold.net/Autonomy/index.php/autonomy/article/view/OP1/pdf (accessed November 2016).

Singer, J. 1999. "Why Can't You Be Normal for Once in Your Life? From a Problem with No Name to the Emergency of a New Category of Difference." In M. Corker and S. French, eds., *Disability Discourse.* Buckingham: Open University Press. Pp. 59–67.

Singh, I., and N. Rose. 2009. "Biomarkers in Psychiatry." *Nature* 460: 202–7.

Singh, J. S. 2016. *Multiple Autisms: Spectrums of Advocacy and Genomic Science.* Minneapolis: University of Minnesota Press.

Slaby, J., and S. Choudhury. 2011. "Proposal for a Critical Neuroscience" In S. Choudhury and J. Slaby, eds., *Critical Neuroscience.* Wiley-Blackwell. Available at http://onlinelibrary.wiley.com/doi/10.1002/9781444343359.ch1/summary (accessed March 25, 2016).

Spencer, M. D., R. J. Holt, L. R. Chura, J. Suckling, A. J. Calder, E. T. Bullmore, and S. Baron-Cohen. 2011. "A Novel Functional Brain Imaging Endophenotype of Autism: The Neural Response to Facial Expression of Emotion." *Translational Psychiatry* 1: e19.

Srivastava, S. 2016. "Everything Is Fucked: The Syllabus." *The Hardest Science* (blog). Available at https://hardsci.wordpress.com/2016/08/11/everything-is-fucked-the-syllabus/ (accessed August 19, 2016).

Stengers, I. 2011. *Thinking with Whitehead: A Free and Wild Creation of Concepts.* Cambridge, MA: Harvard University Press.

Trimble, P.M.R. 1996. *Biological Psychiatry.* London: Wiley-Blackwell.

Tutton, R. 2011. "Promising Pessimism: Reading the Futures To Be Avoided in Biotech." *Social Studies of Science* 41 (3): 411–29.

van Lente, H. 2000. "Forceful Futures: From Promise to Requirement." In N. Brown, B. Rappert, and A. Webster, eds., *Contested Futures: A Sociology of Prospective Techno-Science.* Aldershot: Ashgate. Pp. 43–64.

van Lente, H., and A. Rip. 1998. "Expectations in Technological Developments: An Example of Prospective Structures To Be Filled by Agency." In C. Disco and B. ven der Meulen, eds., *Getting New Technologies Together: Studies in Making Sociotechnical Order.* Berlin: De Gruyter. Pp. 203–31.

Vrecko, S. 2010. "Birth of a Brain Disease: Science, the State, and Addiction Neuropolitics." *History of the Human Sciences* 23 (4): 52–67.

Vul, E., C. Harris, P. Winkielman, and H. Pashler. 2009. "Puzzlingly High Correlations in fMRI Studies of Emotion, Personality, and Social Cognition." *Perspectives on Psychological Science* 4 (3): 274–90.

Wakefield, A., S. Murch, A. Anthony, J. Linnell, D. Casson, M. Malik, M. Berelowitz, A. Dhillon, M. Thomson, P. Harvey, A. Valentine, S. Davies, and J. Walker-

Smith. 1998. "Ileal-Lymphoid-Nodular Hyperplasia, Non-Specific Colitis, and Pervasive Developmental Disorder in Children." *The Lancet* 351 (9103): 637–41.

Walsh, P., M. Elsabbagh, P. Bolton, and I. Singh. 2011. "In Search of Biomarkers for Autism: Scientific, Social, and Ethical Challenges." *Nature Reviews Neuroscience* 12 (10): 603–12.

Weaver, I.C.G., N. Cervoni, F. A. Champagne, A. C. D'Alessio, S. Sharma, J. R. Seckl, S. Dymov, M. Szyf, and M. J. Meaney. 2004. "Epigenetic Programming by Maternal Behavior." *Nature Neuroscience* 7 (8): 847–54.

Weber, M. 1919. "Science as a Vocation." Available at http://anthropos-lab.net/wp/wp-content/uploads/2011/12/Weber-Science-as-a-Vocation.pdf (accessed November 4, 2016).

Weintraub, K. 2015. "Yale Searching for More Objective Way to Diagnose Autism." *Hartford Courant*. Available at www.courant.com/health/hc-yale-autism-study-20151207-story.html (accessed March 21, 2016).

Weisberg, D. S., F. C. Keil, J. Goodstein, E. Rawson, and J. R. Gray. 2008. "The Seductive Allure of Neuroscience Explanations." *Journal of Cognitive Neuroscience* 20 (3): 470–77.

Werling, D. M., and D. H. Geschwind. 2013. "Sex Differences in Autism Spectrum Disorders." *Current Opinion in Neurology* 26 (2): 146–53.

Wertheimer, M. 1987. *A Brief History of Psychology.* New York: Holt, Rinehart and Winston.

Whitehead, A. N. 1935. *Adventures of Ideas.* Cambridge: Cambridge University Press.

———. 1964. *The Concept of Nature.* Cambridge: Cambridge University Press.

———. 1979. *Process and Reality.* London: Macmillan.

Wickelgren, I. 2005. "Autistic Brains out of Synch?." *Science* 308: 1856–58.

Williams, S. J., P. Higgs, and S. Katz. 2011. "Neuroculture, Active Aging and the 'Older Brain': Problems, Promises, and Prospects." *Sociology of Health and Illness* 34 (1): 64–78.

Williams, S. J., S. Katz, and P. Martin. 2012. "Neuroscience and Medicalisation: Sociological Reflections on Memory, Medicine, and the Brain." *Advances in Medical Sociology* 13: 231–54.

Willsey, J., and M. W. State. 2015. "Autism Spectrum Disorders: From Genes to Neurobiology." *Current Opinion in Neurobiology* 30: 92–99.

Willyard, C. 2011. "Men: A Growing Minority?" *gradPSYCH*. Available at www.apa.org/gradpsych/2011/01/cover-men.aspx (accessed March 25, 2016).

Wilson, E. A. 1998. *Neural Geographies: Feminism and the Microstructure of Cognition.* London: Routledge.

———. 2004. *Psychosomatic: Feminism and the Neurological Body.* Durham, NC: Duke University Press.

———. 2010. *Affect and Artificial Intelligence.* Seattle: University of Washington Press.

———. 2015. *Gut Feminism.* Durham, NC: Duke University Press Books.

Wing, L. 1981. "Asperger's Syndrome: A Clinical Account." *Psychological Medicine* 11: 115–29.

Wing, L., S. R. Yeates, L. M. Brierley, and J. Gould. 1976. "The Prevalence of Early Childhood Autism: Comparison of Administrative and Epidemiological Studies." *Psychological Medicine* 6 (1): 89–100.

Wolfe, C., ed. 2003. *Zoontologies: The Question of the Animal.* Minneapolis: University of Minnesota Press.

Woolgar, S., and J. Lezaun. 2013. "The Wrong Bin Bag: A Turn to Ontology in Science and Technology Studies?" *Social Studies of Science* 43 (3): 321–40.

Zimmerman, M., J. I. Mattia, and M. A. Posternak. 2002. "Are Subjects in Pharmacological Treatment Trials of Depression Representative of Patients in Routine Clinical Practice?" *American Journal of Psychiatry* 159 (3): 469–73.

INDEX

CPSIA information can be obtained
at www.ICGtesting.com
Printed in the USA
BVOW08s0231070617
486240BV00002B/3/P

9 780295 741918